PAUL DAY

WHOOPING COUGH

A Journey Towards Truth

NSW, Australia
November 2008

ISBN   978-1-4092-0538-8

CONTENTS

INTRODUCTION

I started my adult life as a staunchly loyal supporter of vaccination. Then my wife and I started having children. Apart from the turmoil and overwhelming sense of responsibility that comes with parenthood, it is also a time of questioning. One of these questions is always, "What are the right health choices for my child?"

I know when I asked this question for the first time I felt singularly incompetent but mustered the resolve to start somewhere. No sooner had my first child entered this world (in a way I must say was unexpectedly graphic as it was overwhelming) than the question of vaccination was asked (almost immediately).

A friend of ours cast a shadow of doubt over the benefits of vaccinations generally so I thought this might be a good place to start my quest for the perfect child. After all, at that time, health professionals had always discussed vaccinations and requested notional consent in a way that suggested that I was taking an informed and competent position on the issue. While I knew I had portrayed understanding and competence in a fraudulent way, I thought that it was such a big issue that there would be plenty of information to support the confident position of these health professionals. I thought the question of vaccinations would be low hanging fruit on the tree of "What are the right health choices for my child?" knowledge.

That was four children and thirteen years ago! While this vexing issue has not distracted me completely from discovering other important things about my children (they are not yet perfect in everyone's eyes but this is simply a matter of time) it has been singularly frustrating. I have been astounded at how difficult it has been to collect clear and concise information on vaccines and how emotive this issue seems to be.

While I have examined all the vaccinations in Australia with varying intensity, more recently, I decided to concentrate my efforts on the pertussis (whooping cough) vaccine. Pertussis seems to be at the forefront of most vaccine debates due, in part, to the graphic health department advertisements on television, the fairly disturbing symptoms of the disease in its worst form and the relatively high mortality rate compared to other diseases covered by the Australian immunisation schedule.

I have listed the 15 simple questions that I asked myself at the beginning of my research (sections 2 to 16 inclusive). At the end of each section/question I have summarised my position. I then present these summaries together in section 17.

I have written this book as support for my position on the question of whether to vaccinate against the pertussis disease and also as an offering for anybody interested. I make no excuses for the writing style because …. I haven't got any and in any case this information is mostly based on pear reviewed and respected medical journals so, if anything, the style tends toward the technical. I am not a medical professional and do not represent myself as anything other then a genuinely concerned parent searching for the best health outcome for my children.

2

WHAT IS PERTUSSIS (COMMONLY KNOWN AS WHOOPING COUGH)?

The information in this section is taken from a number of reports including (32), (33). This is near the beginning of the book and a wise person once told me never to begin something important with an apology. Well, this is my first and last apology; sorry this bit is tricky to read. I seriously considered leaving it until later in the book but I think the book needed the foundation. Anyway …….. here it is ….. please refer to the glossary if that helps.

2.1 The Organism

Bordetella pertussis is a fastidious Gram negative bacterium that colonizes the respiratory epithelium. Adherence to epithelial cells of the respiratory tract involves bacterial surface antigens, in particular filamentous haemagglutinin (FHA). The bacterium then produces a series of active toxins, including the pertussis toxin and other neurologically active toxins, which together produce the clinical symptoms described in 2.2 below. The bacterium does not invade tissues and therefore does not cause damage in the same way as the respiratory organisms that may cause classical pneumonia.

B. pertussis serotype is determined by the fimbrial antigens (agglutinogens 1, 2 and 3) expressed on the surface of the organism, giving rise to three serotypes: type 1, 2, 3; type 1, 2; and type 1, 3. Serotypes 1,2 and 1,3 are predominant in clinical isolates in Australia.

The serotype prevalence may be affected by the immunization rates in the community. Serotype 1,2 appears to predominate in unvaccinated communities. Its use in early vaccines led to emergence of type 1,3 as the predominant serotype in many countries. In Australia studies indicate a shift from predominant serotype 1,2 to that of 1,3 from 1987 onwards.

2.2   Clinical Symptoms

The incubation period for pertussis is 6 to 20 days, but usually less than 14 days.

The clinical phases of the disease are:

i.   the prodromal/incubation phase
ii.   the catarrhal phase which lasts about one to two weeks and is characterized by non-specific symptoms that mimic a cold or other viral upper respiratory tract infection, i.e. rhinorrhoea, sneezing, mild cough increasing in intensity, conjunctivitis and malaise with little or no fever. During this phase the disease becomes increasingly contagious;
iii. the paroxysmal phase that is characterized by sudden attacks of severe, repetitive coughing culminating in the characteristic inspiratory whoop often followed by vomiting. Transmission is most efficient in this stage of the disease, which last up to four weeks; and
iv.   the convalescent phase which is marked by a reduction in the frequency and duration of coughing spells. Complete recovery may take from several weeks to three or more months.

2.3   Clinical Diagnosis

Clinical diagnosis based on the duration of cough and the characteristic nature of the disease is critical in identifying pertussis because there are difficulties in obtaining laboratory confirmation of the diagnosis.

Clinical diagnosis based on the presence of coughing paroxysms, inspiratory "whoop" or post-tussive vomiting without other apparent cause is often sufficiently distinctive to identify pertussis. Infants less than six months of age, and to a lesser extent those less than 12 months of age, are less likely to present with paroxysmal cough. For these children, severe apnoea may be the only or dominant manifestation.

The classical picture of whooping cough as typified by a moderately ill child with paroxysmal cough and characteristic whoop *is not* often encountered, though a persistent cough with vomiting *is* common (10).

2.4   Laboratory Diagnosis

Specific issues relating to laboratory diagnosis are as follows:

i.   Serology

Serologic testing of pertussis has not been standardized and should not be considered as a definitive diagnosis. Where serologic testing is requested, IgA serology should be specified. However, Bordetella pertussis - specific IgA is less sensitive in children under two years of age, and is of little value in children less than three months of age.

ii.   Culture

At present, culture is the only standardized test for laboratory confirmation of pertussis and should be performed in a laboratory with established expertise in isolating the organism. Nasopharyngeal cultures should be collected by either aspiration (suction device) or using a calcium alginate swab (cotton swabs must not be used). There is no role for oropharyngeal culture as results are likely to be falsely negative.

10

iii.   PCR

Identification of Bordetella pertussis in nasopharyngeal secretions by the polymerase chain reaction (PCR) technique is in the development stage but should be considered as a confirmed diagnosis if undertaken in a laboratory with established expertise in the area.

iv.   DFA

Owing to an unacceptably high rate of false-positive and false-negative results, commercially available direct fluorescent antibody (DFA) tests are of no use.

2.5   Case Definitions

i.   Probable

    a.   A cough illness lasting 14 days or more with one of the following: paroxysms of coughing; inspiratory "whoop" or post-tussive vomiting, without other apparent cause; or

    b.   A cough illness lasting 14 days or more in a patient with Bordetalla pertussis – specific IgA detected serum.

ii.   Laboratory Confirmed

    a.   Isolation of Bordetella pertussis from a clinical specimen; or

    b.   Positive polymerase chain reaction (PCR) assay for Bordetalla pertussis undertaken in a laboratory with established expertise in the area.

iii.   Epidemiologically Confirmed

A cough illness lasting 14 days or more in a patient who is epidemiologically linked to a laboratory confirmed case.

My Position (What is Pertussis?):

Pertussis is a long disease that can be distressing for the patient and carer particularly during the paroxysmal phase. Laboratory confirmation of the disease is best achieved via a nasopharyngeal swab that can also be distressing for a child. Epidemiological confirmation of existing symptoms can be enough to commence treatment.

3

HOW IS PERTUSSIS TRANSMITTED?

The information in this section is taken from a number of reports including (32), (33).

Bordetella pertussis is highly infectious. It may be spread from person to person by close contact usually by respiratory aerosols, infecting 70 to 100 per cent of household contacts.

Although the definition of significant contact may vary according to the situation, direct contact with respiratory secretions from a case (e.g. an explosive cough or sneeze, sharing food, sharing utensils during a meal, kissing, mouth-to-mouth resuscitation or full medical/dental examination of the nose, throat or mouth) are generally considered to be significant. Identifying significant contact must be on an individual basis and take into consideration the degree of risk to the individual and the specifics of the exposure. For example, a significant contact may include an infant being in the same room for an hour with a case or a newborn being directly coughed upon by a case.

Contact within a family day-care setting is considered equivalent to household contact. In day-care centres and preschool, the degree of contact should be assessed according to the individual circumstances and discussed, if necessary, with an experienced public health professional. Where there are morning and afternoon sessions, only the effected group need be regarded as at increased risk; the other group should be given an advisory letter.

While listing specific settings and durations of exposure can be problematic, the following priority for action, in decreasing order, can be used as a guide:

Setting	Persons considered significant contacts
1. Family	All immediate family and household members in contact with the case since onset of the disease, usually taken as the date of cough onset. Very close friends with whom there was generally at least one hour of close contact a day should be included.
2. Family day-care	All residents of the household providing care and all attendees in contact with the case since onset of the disease.

3. Day-care and Preschool

<u>One Case</u>
All children under one year of age and children not up-to-date with the recommended vaccinations against pertussis who have been in contact with the case since onset of the disease.
<u>> One Case</u>
All attendees and staff in contact with the cases since onset of the disease.

4. Closed settings (e.g. boarding house)

All individuals who slept in the same room, dormitory or tent with the case.

My Position (How is Pertussis transmitted?):
Bordetella pertussis is highly infectious. It may be spread from person to person by close contact usually by respiratory aerosols. Identifying a significant contact must be on an individual basis and take into consideration the degree of risk to the individual and the specifics of the exposure.

WHAT IS THE INCIDENCE OF PERTUSSIS AND HOW DOES IT VARY WITH AGE?

The table below shows a summary of the incidence of pertussis in Australia by age taken from (32). The averages are taken over a 5-year period between 1993 and 1998. This is a good sample because it includes a 4 yearly peak although immunisation rates probably varied over this period. Assuming an Australian population of 17,000,000 the average notification rate during this period was 35/100,000

This information is a bit misleading because Australia did sustain a much lower notification rate of between 2 and 5/100,000 between 1968 and 1992. This figure is more in line with benchmarks such as the USA and the Netherlands (refer section 8.1), although both of these countries have experienced an unexplained resurgence in the notification rate (5), (6). Also, it should be noted that vaccination uptake and the Australian immunisation schedule varied significantly between 1968 and 1992 (refer section 10.2) without a noticeable change in notification rate. This suggests that notification rates are a function of more than just vaccination rates (refer to section 8.1 for more on this).

Then in 1993-94 something happened that meant that the notification rate increased tenfold where it has stayed. No-one is really sure what happened and explanations include: poor vaccination coverage; waning immunity; the introduction of laboratory reporting in some States and Territories; increased awareness of pertussis and the requirement to notify cases; and/or increased testing, particularly serological testing (33).

AVERAGE PERTUSSIS INCIDENCE PER YEAR

AGE	NOTIFICATIONS	%	HOSPITALISATION	%	DEATH
0 TO 1**	460	8%	460	48%	1.8
1 TO 4	460	8%	226	23%	0
5 TO 9	1128	19%	74	8%	0
10 TO 14	1128	19%	74	8%	0
15 TO 24	629	11%	22	2%	0
25 TO 59	1706	29%	69	7%	0
60+	339	6%	37	4%	0
TOTAL	5848	100%	961	100%	0

** Notifications = Hospitalisations (Actual average notifications = 345)

There is one important point that must be considered while examining these numbers and that is they are based on notifications. It is widely accepted that the actual incidence of pertussis could be as much as 10 times greater than the notification rate. An English study (36) that examined a community over 10-years showed that the actual incidence of pertussis was 8 times greater than the notification rate at the time. This could be due to a host of reasons including incorrect diagnosis, doctors unaware of notification requirements, lack of public health resources, etc. Refer to Section 8.3 for more information in this area.

There are approximately 250,000 live births each year in Australia. The risk of;

1.   a newborn contracting pertussis within the first year is between 0.2% (based on health department notifications) and 2% (based on implicit information from more detailed community studies).

2.   a newborn dying from pertussis within the first year is 0.00072% (or 7.2 in a million).

3.   a newborn dying from pertussis within the first year after hospital admission is 0.4% (based on health department notifications).

16

The total notification rate of Pertussis increases every 3-4 years, shown below. The media (and some doctors) call these epidemics when these cycles have been known to occur regularly. In America the cycles still occur even with an immunisation rate of over 95%.

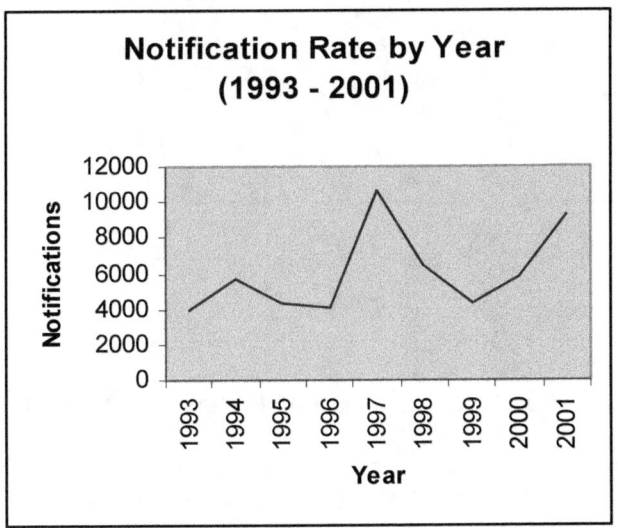

In Australia the disease exhibits some seasonality, the number of notifications increasing in summer (by as much as 3 times). Vaccine uptake does seem to have some impact on smoothing this seasonal effect.

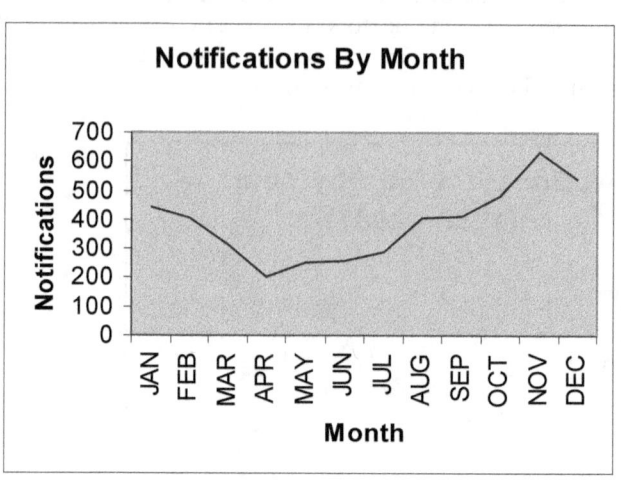

My Position (What is the incidence of Pertussis in Australia and how does it vary with age?):

The incidence of pertussis in Australia is anywhere between 35 and 350/100,000 depending on the way the incidence is assessed. The incidence increases up to threefold in the summer months and yearly rates double every 4 years.

Rates of notification and hospitalisation are much higher (as high as 2% for children under 1 year of age) for infants and young children.

5

HOW SEVERE IS THE DISEASE AND WHAT IS THE EFFECT OF VACCINES ON SEVERITY?

5.1   Disease Severity

The severity of a case of whooping cough can be expressed in 3 ways - that is, the duration of the illness, the maximum number of coughing spasms per 24 hours, and the incidence of complications. There are only a handful of studies on this subject with (because of its nature) not enough of a sample to make the results conclusive.

The disease itself is a long one lasting, on average, about 50 days from the onset of the disease until the child is completely recovered (10). The duration of the disease, on average, seems to be reduced by about 10-30% when vaccinated children are compared to unvaccinated. In children admitted to hospital, however, vaccination against whooping cough does not modify the disease if the total duration of the patient's disease is used as an index of severity (11). The duration reduces by another 10% for children over 1 year of age (11).

At the peak of his/her illness a child will experience about 12 coughing spasms in a 24-hour period. Again this seems to be reduced by 10-30% with vaccination.

19

About 17% of people who contract whooping cough will endure complications, 50% of these will be minor such as;

       i.   Conjunctival Haemorrage

      ii.   Ulceration of the Frenum.

The rest will be major such as;

       i.   Apnoea with or without Cyanosis

      ii.   Bronchitis

     iii.   Pneumonia

     iv.   Severe Dehydration

      v.   Convulsions.

In Australia about 16-17% of the people notified as contracting whooping cough will be admitted to hospital (which varies depending on variables such as accessibility of hospitals or the rate at which cases are notified relative to the actual incidence). The mean length of stay in hospital is 2 to 4 days. Of those admitted to hospital approximately;

i. 70-90% are under 5 years
ii. 50% are aged less than 12 months, and
iii. 15% are aged less than 3 months which reflects the increased severity of the disease in the very young.
iv. 20% will get pneumonia.
v. 7% will be admitted to intensive care.
vi. 4% will require artificial ventilation (13) (of those requiring ventilation over 90% will be under 6 months of age).
vii. 0.3 to 1.4% will endure encephalopathy (refer to section 10.2)
viii. 0.5% will experience generalized fitting.
ix. 0.5% will die.

Umbilical and inguinal hernias are often exaggerated by the cough. Rectal prolapse is a rare complication that occurs in malnourished patients.

5.2   Mortality

It seems that Whooping Cough vaccination has been least effective in terms of mortality. Calculations based on the mortality of whooping cough before 1957 (*and before vaccination*) predict accurately the subsequent decline and the present low mortality.(12) It appears that the main factor which has reduced the mortality of this disease has been the routine use of chemotherapeutic agents to prevent and treat pneumonia.(11) Refer to Section 8.2 for more information.

While nearly all patients that die from Pertussis will have contracted pneumonia, death will occur from respiratory failure, usually from apnoea rather than from secondary infection or aspiration.(13)(28)

In Australia, from October 1996 to September 1997, seven children between two weeks and four months of age died from pertussis. In the preceding 20 years, from 1976 to 1995, 21 children died from pertussis. All but one of these was less than one year of age. (33)

A report released in 1996 on pertussis deaths in the USA (28) identified patients who died of pertussis whose dates of disease onset were between January 1992 and December 1993 (see chart on next page). Of the 23 reported deaths attributable to pertussis, 20 (87%) were younger than 9 months of age and 16 (70%) were less than 3 months old. Of the 3 children that were older than nine months, one (14 months of age) was born at 26 weeks' gestation and the other two (age 8 and 9 years) were not vaccinated because they had severe neurological disorders, and it had been thought that pertussis vaccine was contraindicated.

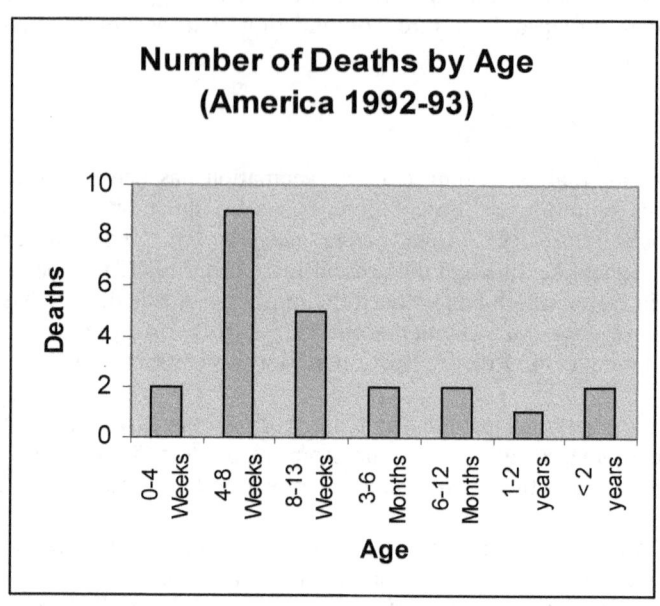

The American report also indicated that risk factors for pertussis deaths included a gestation period of less than 36 weeks. Among the 20 children for whom gestational ages were known, 12 (60%) were born at 36 weeks' gestation or earlier; in contrast, 10.7% of live births in the United States in 1992 were at 36 weeks' gestation or earlier. The same report went on to say that premature infants have previously been reported to be overrepresented in two reports of hospitalised cases of pertussis, suggesting that prematurity may be a risk factor for pertussis-related pneumonia and death.

As for the mortality rate, about 0.3-0.5% of hospital admissions due to whooping cough end in death. Mortality seems to be lower in an environment of low vaccine uptake such as Sweden where about 0.1% of hospital admissions die. (7)(13)

My Position (How severe is the disease and what is the effect of vaccines on severity?):

In its worst form pertussis can be a frightening disease, particularly in the young. The severity of a case of whooping cough can be expressed in 3 ways, the duration of the illness, the maximum number of coughing spasms per 24hours, and the incidence of complications.

The disease usually lasts 50 days travelling through different phases of severity. At the peak of his/her illness a child will experience about 12 coughing spasms in a 24-hour period. About 16% of the people notified as contracting whooping cough in Australia will be admitted to hospital. Of those admitted to hospital about 70-90% are under 5 years, 50% are aged less than 12 months and 15% less than 3 months which reflects the increased severity of the disease in the very young.

About 20% of hospital admissions will get pneumonia and perhaps 0.5% will experience generalized fitting. 7% of the hospital admissions will be admitted to intensive care, 4% will require artificial ventilation (13) (of those requiring ventilation over 90% will be under 6 months of age).

0.3-0.5% of hospital admissions in Australia due to whooping cough end in death. While nearly all patients that die from Pertussis will have contracted pneumonia, death will occur from respiratory failure, usually from apnoea rather than from secondary infection or aspiration. Of those that die from complications relating to whooping cough, nearly all will be younger than 9 months. The evidence suggests that those older than 9 months will have other longer term health issues. It seems that prematurity may be a risk factor for pertussis-related pneumonia and death.

The evidence also suggests that vaccination reduces the main severity indicators (i.e. disease duration, coughing spasms and complications) by 10-30%.

6

HOW EFFECTIVE IS THE VACCINE IN TERMS OF PROTECTION?

6.1   How is Vaccine Effectiveness Measured?

The efficacy of a vaccine is generally measured as a percentage using the following formula;

$$= \frac{\text{(attack rate in unimmunised group} - \text{attack rate in immunised group)} \times 100}{\text{(attack rate in immunised group)}}$$

6.2   What is the History of Vaccines?

The following information is taken from (34):

      i.   Whole cell vaccines

The first whole-cell vaccines in Australia were made by the then Commonwealth Serum Laboratories (now CSL Limited) in about 1920, but were not used widely until the 1940s. At this time pertussis, diphtheria and tetanus vaccines still had to be given as separate injections, and debate began about whether it was possible to combine antigens. By 1953 the first Australian-made Triple Antigen (DTPw) (diphtheria and tetanus toxoids with whole-cell pertussis) was produced.

ii.   Acellular vaccines

Japanese investigators accelerated the development of acellular pertussis vaccines in the 1970s. This followed an epidemic of pertussis that occurred after the cessation of whole-cell pertussis immunisation, in early 1971, because of concern about adverse effects. Development of the acellular vaccines became possible once biologically active and extractable components of Bordetella pertussis were identified. One or more of the following five components are included in all acellular vaccines developed to date:

        a.   detoxified pertussis toxin (PT);
        b.   the outer membrane protein pertactin (PRN); and
        c.   three surface proteins – filamentous haemagglutinin (FHA) and two agglutinogens (AGGs).

The first acellular vaccines were strongly influenced by the notion that pertussis was a single toxin disease, like diphtheria, and could be prevented by use of a pertussis toxoid. This is incorrect, partly because Bordetella parapertussis, which does not produce pertussis toxin, causes an almost identical clinical picture.

Around 1997 a 3-component vaccine (Infanrix, SmithKline Beecham) was approved for marketing (licensed) in Australia. A 5-component vaccine (Tripacel, CSL Vaccines, manufactured by Connaught Laboratories, Canada) has been licensed. Both vaccines are approved for use for primary and booster doses.

The acellular vaccines are approximately four times the cost of whole-cell vaccines (34)

6.3 How well does the vaccine protect against the disease?

i. Whole cell vaccines

A number of candidate vaccines were examined in trials conducted by the Medical Research Council in the United Kingdom in the 1950s. These trials established a correlation between clinical efficacy and the mouse protection test (Kendrick assay), which has been used ever since to monitor potency of whole-cell vaccines. Australia has adopted the United Kingdom's criterion of requiring 4 mouse protection international units (IU), but the United States of America has allowed vaccines to have as low as 2 IU. One of the outcomes of the recent comparative trials has been evidence that whole-cell vaccines may vary significantly in efficacy (see table below). It is suggested that whole-cell vaccines, such as the CSL vaccine, which pass the more stringent mouse protection test (4IU) are likely to be more protective, but there are limited observational (household contact) data and no trial data estimating the efficacy of the Australian whole-cell vaccine. Some experts believe that if the vaccine contains all 3 agglutinogens (1,2,3) it is more likely to be protective against all 3 serotypes of the organism (type 1,2,3; type 1,2; and type 1,3).

ii. Acellular vaccines

In contrast to whole-cell vaccines, the Kendrick test does not correlate efficacy for acellular vaccines, making large trials the only means of evaluating efficacy. Eight large controlled trials have now been published, all but one in Europe, to evaluate the efficacy of the acellular vaccines. Differences in methodology and case definitions make comparisons between trials difficult. In particular, the World Health Organization (WHO) case definition (21 days of cough) detects only typical whooping cough, which is more common in unvaccinated individuals. Using the WHO definition therefore inflates vaccine efficacy estimates compared with case definitions which include milder but still culture positive infections. When mild cases are taken into account, the efficacy of acellular vaccines varies widely.

Their efficacy broadly correlates with increasing numbers of components, from around 50% with 1 or 2 components to around 70% with at least 3 components, including PRN, and 80% or more with 5 component vaccines. In the Swedish trial, a 5-component acellular vaccine and the British whole-cell vaccine gave better protection against less severe disease (laboratory confirmed pertussis with or without cough, and whooping cough diagnosed by the child's parents) than a 3-component vaccine (not the 3-component vaccine currently approved in Australia).

Site	Composition				Schedule	Efficacy (95%CI)	
	PT	FHA	PRN	FIM	(months)	DTPa	DTPw
Germany	x	x			2, 4, 6	96 (78-99)2	97 (79-100)
Sweden	x	x	x	x	2, 4, 6	85 (81-89)	48 (37-58)3
Sweden	x	x			2, 4, 6	59 (51-66)	48 (14-52)3
Italy	x	x	x		2, 4, 6	84 (76-90)	36 (14-52)3
Italy	x	x	x		2, 4, 6	84 (76-90)	36 (14-52)3
Sweden	x				2, 5, 12	71 (63-78)2	Not Tested
Germany	x	x	x		3, 4, 5	89 (77-95)	97 (83-100)
Africa	x	x			2, 4, 6	86 2	96

1. Using WHO definition of 21 days or more cough

2. These results are likely to be too high due to study methods and observer bias

3. Study used Connaught (Canada) whole-cell vaccine.

DTPa = Acellular Diphtheria – Tetanus – Pertussis vaccine.

DTPw = Whole-cell Diphtheria – Tetanus – Pertussis vaccine

6.4   How does immunity from the vaccine vary with time

There is only one report (that I have discovered) published in the British Medical Journal in 1988 that examines the duration of effectiveness of pertussis vaccine (36). This was a very ambitious study undertaken in a general practice community. The study monitored the occurrence of whooping cough in children aged between 1-7 years of age within the community over a 10-year period from 1977 to 1988. Only whole cell vaccines were approved for use in the United Kingdom during this study. The study allowed for variations in vaccine uptake.

The table below shows the findings. These contradict the studies described in section 6.3 and this may be due to a number of factors including differences in the vaccine, method of diagnosis, knowledge of previous disease, numbers of missed cases, and other unrecognised factors.

Efficacy of pertussis vaccine in children by age							
Age (Years)	1	2	3	4	5	6	7
Efficacy of vaccine (%)	100	96	89	84	52	54	46

This study was one of the main catalysts behind including a booster shot of DTP vaccine for children entering school at 5 years of age. Vaccine efficacy percentages shown in this study should be used cautiously given the differences when compared to the information in 6.3. However, it certainly shows that efficacy diminishes significantly after 5 years. Consequently, if boosters are given to children at 5 years then we can assume that vaccine efficacy will follow the chart shown below (although this is my extrapolation and has not been proven).

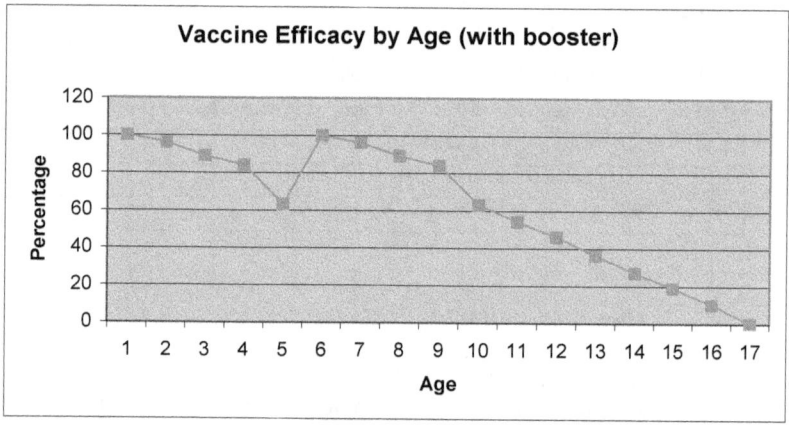

This is the best case performance of any pertussis vaccine. The table in 6.3 clearly shows that the efficacy of vaccines vary widely depending on many factors including manufacturer and type of vaccine. Consequently all we know for sure is that the efficacy of any vaccine falls somewhere within a band below this curve. To find the bottom of the band for given vaccines subtract the following percentages from any point on the efficacy line shown in the graph above.

Whole cell vaccines	Subtract 30%
1 or 2 component acellular vaccines	Subtract 50%
3 or 4 component acellular vaccines	Subtract 30%
5 component acellular vaccines	Subtract 20%

It is interesting to note that none of the efficacy studies monitored children before their vaccination schedule was completed. Therefore I could find no credible data on the increasing immunity to pertussis with each shot. The current Australian schedule recommends pertussis shots at 2, 4, and 6 months and then boosters at 18 months and 4-5 years. So for the purposes of most studies (including this one) the protective effect of the pertussis vaccine can be assumed 0% until the child is 6 months old.

Another observation that I believe is worth mentioning concerns the age at which the efficacy of the vaccine reduces to zero. This happens when the child reaches the age of approximately 15-17 years. Referring to the table in Section 4 this age happens to coincide with the age at which the pertussis hospitalisation rate drops to 2% from 8%-48% for other, younger age groups. The cynic could argue that the government promotes vaccination only to the extent that it minimises hospital admissions.

My Position (How effective is the vaccine in terms of protection?):
The effectiveness (efficacy) of vaccines can vary anywhere between 35% and 97% depending on the type of vaccine and which of the 3 serotypes of the organism are prevalent at the time (type 1,2,3; type 1,2; and type 1,3).

Acellular vaccines are proving to be more effective than whole cell vaccines. There is no specific data on the 3-component acellular vaccine used in Australia however there is enough evidence to suggest that it's efficacy in the first couple of years after the final shot would lie between 70% and 97%.

In the absence of other data it can be assumed that the protective effect of acellular vaccines over time is similar to whole cell vaccines. In which case the protective effect will have reduced to 0% by the time a child reaches his/her mid to late teens.

Also, I could find no information to support the notion that the protective effect of vaccines is anything other than 0% before the final shot is administered at the age of 6 months. My belief is that this is probably too conservative.

WHAT IS THE PURPOSE
OF THE VACCINE?

The Pertussis Working Party convened in 1996 by the National Health and Medical Research Council (NHMRC) claimed that strategies to reduce pertussis needed to concentrate on:

i.   achieving very high (95 per cent or more) vaccination coverage;
ii. introducing vaccines with fewer side effects which would be more acceptable to the public;
iii. improving outbreak control to reduce transmission of the disease; and
iv. increasing public awareness and provider commitment to immunisation.(33)

Really this can be distilled into only two strategies:

i.   increase vaccine coverage
ii. improve outbreak control to reduce transmission of the disease.

If the pertussis vaccine where 100 per cent effective and given at the optimum time, it has been estimated that between 92 and 95 per cent of the population need to be immunised to block transmission of pertussis and provide conditions generally required for herd immunity (33). Given:

    i.   the efficacy of the vaccines (i.e. not 100%);
    ii.   the fact that immunity to pertussis due to the vaccines reduces with time; and
    iii.   the reasonably complex vaccine schedule;

I don't think either condition for herd immunity can be met.

Consequently, the primary goal of the pertussis vaccine must be to reduce exposure of infants (particularly those under one year of age) to the disease. Then once exposed, the vaccine offers one last barrier in terms of protection against infection. I have dealt with the question of exposure in section 7.1 and then with protection against infection in section 14.

7.1   How do Vaccines reduce the risk of exposure for a Child?

The time of greatest concern for our family was in 2002 when, in our household, there were twins under the age of one, one child 4 years, one child of 7 years and two adults. My main concern was for the twins (i.e. in relation to pertussis) as the disease would be the most severe for them, possibly in the worst case causing death. The riskiest environment for them was if we sent them to childcare and did not vaccinate their siblings.

I have shown the statistical picture that would be present each year below, assuming that the protective effect of the vaccine that my wife and I received as children has worn off and noting that all children are unimmunised. These probabilities represent my estimate of actual incidence in Australia as distinct from notification rate (this distinction is dealt with in more detail as part of section 8.3).

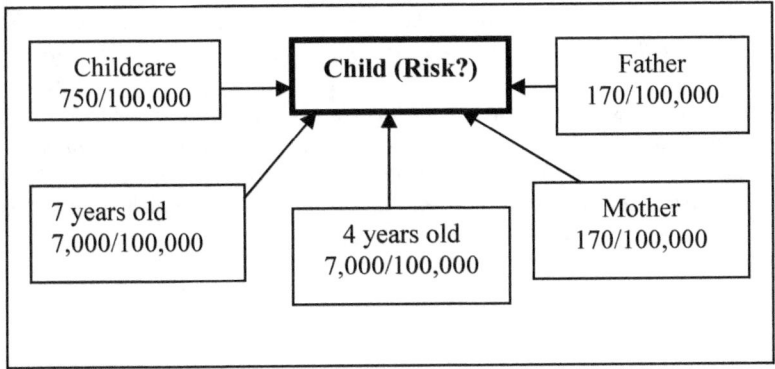

The dynamics of pertussis transmission, including mother to infant transmission, is unknown. However it would seem logical that while immunising the siblings would reduce the risk to the child, there is still a significantly high exposure risk from the close contact afforded by the mother and father as well as childcare.

If simple precautions are taken while siblings are at home then the chances of the child under one contracting the disease before symptoms appear in the other family members are minimized. If however the child goes to childcare or family daycare regularly then the chances of the child being involved in a high-risk contact and therefore being the first to contract the disease without warning increases markedly.

If the child does not go to family daycare and the siblings, mother or father contract pertussis then precautions can be taken to minimize the impact of the disease on other household contacts (see section 14). In fact in some ways it is better for the other siblings to be unimmunised so that the symptoms are more obvious and prophylactic antibiotics may be administered straight away.

My Position (What is the purpose of the vaccine?):
The pertussis vaccine will never provide herd immunity because it is not 100% effective and its protective effect reduces significantly with time. Children under 1 year of age are the most at risk should they contract Pertussis however, the protective effect of the vaccine is considered 0 until all shots are completed at 6 months.

Consequently, the primary goal of the pertussis vaccine is to reduce exposure of young children (particularly those under 1 year of age) to the disease. Then if exposed, the vaccine offers the children (older than 6 months) one last barrier in terms of protection against infection.

The benefits of immunising siblings within a family in order to reduce the exposure risk to a child under one year old (therefore minimising the pertussis related death rate) is questionable considering the child's high risk exposure to it's mother and father who are essentially unimmunised. More research is required in this area.

WHAT IS THE EFFECT OF VACCINE COVERAGE ON THE INCIDENCE OF PERTUSSIS?

8.1 Notifications vs. Vaccine Coverage

We have to look at the experience of different countries to get at least a broad indication of how notification rates vary with vaccine coverage. I have assembled a table below that draws from a series of references including (8), (7), (33), (34), (36) and (37). These figures do not necessarily represent current vaccination rates in each country, however these countries have reached the nominated vaccine coverage rates at some time and the corresponding notification rates have been documented.

Country	Vaccine Description	Estimated Vaccine Coverage	Nominal Notification Rate Total (per 100,000)
Sweden	None	<10%	60 – 130 (say 90)
England & Wales	Whole Cell	30%	50 – 100 (say 75)
Russia	Whole Cell	60-70%	22
Australia	Whole Cell & Acellular	60-75%	35
USA/Netherlands	Whole Cell & Acellular	90-95%	2
Poland	N.A.	95%	1
Hungary	Whole Cell	100%	<1

35

So representing this information graphically we get a nominal curve similar to the graph shown below.

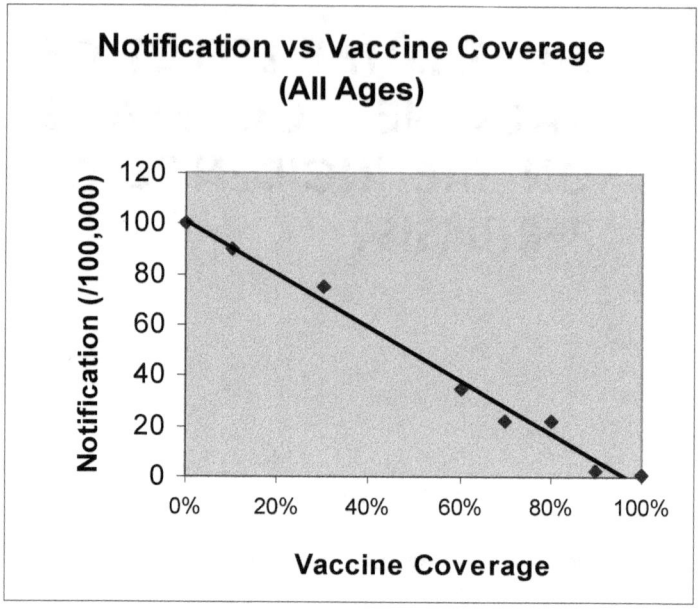

It seems that the relationship between vaccine coverage and notification rates is linear. If there was a herd effect caused by vaccine uptake then we would expect that some sort of critical mass would be achieved and that the line of best fit in the curve above would be more logarithmic.

If we now take the notification rate for children <12 months we get the table shown below. This information has been compiled from several reports (37), (40), (7), (42) & (32).

Country	Estimated Vaccine Coverage	Nominal Notification Rate < 12 months (per 100,000)
Sweden	<10%	630 – 1000 (say 800)
Australia	60-75%	184
USA/Netherlands	90-95%	40

So representing this information graphically we get a nominal curve shown below.

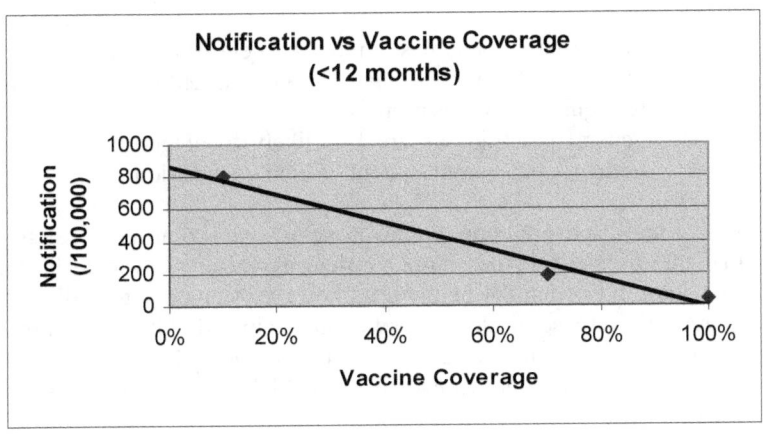

8.2   Deaths vs. Vaccine Coverage

I have compiled the table below from a series of reports (32), (7) and (40).

Country	Estimated Vaccine Coverage	Hospital Admissions as % of Notifications	Deaths as % of Hospital Admissions
Sweden	<10%	6.8	0.1
Australia	60-75%	16.4	0.19
USA/Netherlands	90-95%	42.6	0.86

This is an interesting table because it shows that hospital admissions as a percentage of notifications actually increases steadily with vaccine coverage.

I think hospital admission numbers are more reliable than notification figures and should be used as the base line. This is because hospitals are much more likely to keep accurate records and be relatively vigilant and accurate about a pertussis diagnosis. Also, a hospital admission will always require the general practitioner to consider a diagnosis carefully before recommending admission.

If we sit back then, and consider possible reasons why hospital admissions increase as a percentage of notifications with vaccine coverage, we can really only come up with two hypotheses;

1. the disease becomes more virulent as vaccine coverage increases. So much so that that a confirmed case is more likely to require hospitalisation, or;
2. general practitioners are less likely to diagnose symptoms as pertussis in an environment of high vaccination.

I have not discovered any studies in peer reviewed articles that suggest that the disease becomes more virulent as vaccine coverage increases. The chart in Section 8.4 below does however show a potentially modest increase in disease virulence for the unvaccinated although this trend is inconclusive and, in any case, is nowhere near the fivefold increase shown in the last table. Consequently I think we can confidently reject the first hypothesis.

So what of the second hypothesis? Well, to be fair and consistent, I know of no studies that support this idea. However, I do think it survives test of logic and personal experience. General practitioners use the knowledge of vaccine status as an important aid in determining diagnosis. Hence it would be logical to expect that doctors are less likely to diagnose pertussis in an environment of high vaccine coverage.

Further if we look carefully at the last table we find that <u>deaths</u> as a percentage of hospital admissions also increase with vaccine coverage. Could it be that, in an environment of high vaccine coverage, hospital admission for correctly diagnosed pertussis cases is often delayed until symptoms are more severe and more likely to lead to death. Consequently it seems that there is a slightly perverse effect whereby mortality rates increase in these environments (ie. when compared to hospital admissions).

Finally then, as if to validate the second hypothesis, if we combine all the data in Sections 8.1 and 8.2 we can mathematically derive how death rates vary with vaccine coverage shown on the next page;

Country	Estimated Vaccine Coverage	Deaths (/100,000) All ages
Sweden	<10%	0.009
Australia	60-75%	0.011
USA/Netherlands	90-95%	0.007

So, while it is important not to imply too much precision in these numbers, we can make a broad conclusion. The number of deaths is NOT related to vaccine coverage and that the reduction in death rate over time (particularly as far back as the late 1800's) may have more to do with other factors such as better drugs used in fighting pneumonia and better living conditions.

8.3   Actual Incidence vs. Notification Rate

As shown in section 6.4 the effectiveness of the vaccine deteriorates with time until late teens when there is very little protective effect. Consequently there is always a pool of adults and adolescents which have limited or no resistance to pertussis. There is very little information on the dynamics of pertussis in this part of the community and the notification rate within this demographic varies widely from country to country and is also probably dependant on such factors as living conditions and climate.

If the effect of the vaccine deteriorates over time then this would suggest that the incidence of pertussis in any age group over about 20 years of age would be the same no matter which country was being examined. In fact the incidence in this age group should be reaching those of Sweden, which has <10% vaccine uptake, for all countries.

If we take the USA as an example, the notification rate (not the actual incidence) of pertussis is about 0.1 per 100,000 (40) for people aged <20. A study was undertaken in the USA (43) to identify what the actual incidence of pertussis was in the adult population by examining how many adults with a prolonged cough lasting 2 weeks or longer actually had pertussis. The actual incidence of adult pertussis was estimated to be 176 cases per 100,000 person years (95% confidence interval, 97 to 255 cases) which is closer the Swedish rate and over 1700 times the notification rate!

Hence, the actual incidence of pertussis is much greater than the notification rate which further supports the second hypotheses in Section 8.2. Also, as stated in section 4, for children under the age of 7 years the actual incidence of pertussis is about 8 to 10 times greater than the notification rate. In adults the actual incidence can be as high as 1700 times greater than the notification rate.

Again I think the difference between notification and incidence is caused by a number of factors such as funding to the public health organisation and the willingness of the doctors to report pertussis. Also, I think the discrepancy between notification and incidence rate for children and adults probably reflects the increasing severity of the disease in the young.

Also, in discussing the effect of vaccine coverage on the prevalence of pertussis in a community it is important to recognize that there will always be an underlying pool of people (particularly adults) who are infecting. In fact the cycle of infection and then herd immunity in this group probably lasts about 4 years which goes some way to explaining the 4 yearly endemics that occur with or without high levels of vaccine coverage.

8.4   Actual Attack Rate for Immunised Children vs. Unimmunised Children

If we return to the study which was used to justify the pre-school booster (36), we can extrapolate an actual attack rate (as opposed to notification rate) relative to community immunisation rates for 0-7 year olds shown in the chart below (remembering that this does not take account of the pre school booster shot). The chart shows clearly that attack rates for immunised children do not vary with the immunisation rate within the community. I anything, and this is somewhat inconclusive given the sample size, the only group that presents with an additional risk of pertussis disease is the unimmunised group.

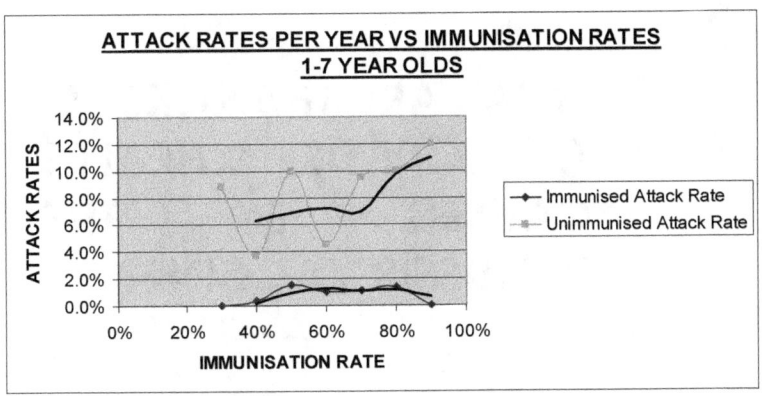

This was interesting for me because I had thought that unimmunised children presented an additional risk to the community. In fact it seems the risk to an immunised child is the same regardless of the rates of immunisation within the community. It also adds credence to the argument that there is always a pool of adults and adolescents who are essentially unimmunised and sustain the disease.

My Position (What is the effect of vaccine coverage on the incidence of pertussis?):

There will always be a pool of unimmunised adults and adolescents regardless of the vaccine coverage unless the longer term protective effect of acellular vaccines is any better than whole cell vaccines. If the acellular vaccines perform similarly to whole cell vaccines then as the coverage rates increase the attack rates for immunised children will not change while the attack rate for unimmunised children will probably increase slightly.

The overall notification rates will reduce in a linear fashion as vaccine coverage increases commensurate with the effectiveness of the vaccine, the prevalent serotype and demographics.

Also considerations such as climate, economic/living conditions, public health budgets, type of disease, case definitions and other yet unknown factors will all have an effect on the absolute minimum notification rate achievable within a given country.

WHAT ARE THE RISKS TO A CHILDS HEALTH CAUSED BY THE PERTUSSIS VACCINE?

9.1 Adverse Events (Reactogenicity)

Whole-cell vaccines contain inactivated B. pertussis organisms and a variable but significant amount of endotoxin, which is probably responsible for the relatively high rate of fever, local reactions, pain and prolonged crying from whole-cell vaccines (see Table below). Endotoxin cannot be exclusively responsible, as these effects also occur, at a lower rate, with DT vaccine. Attempts were made in Australia and elsewhere to eliminate endotoxin from whole cell pertussis vaccines, but this proved difficult and was superseded by development of acellular vaccines. (34)

Lower incidence of the more common but less severe reactions (such as local swelling, pain and fever) with acellular vaccines has .. *apparently* ...been established .... *although I am yet to sight the data/reports*. Data concerning uncommon but more sever reactions, such as fits and HHEs (see section 9.3.2 for a definition of HHE) are more difficult to accumulate, but the combined results of a number of controlled trials ... *apparently* ... show that these are also significantly lower than with whole-cell vaccines and do not appear to be related to the number of components or to any one component. In the United States of America two products were licensed in 1991 for the fourth and fifth infant doses. Surveillance after this licensure showed that post-vaccination seizures and hospitalisation were reduced by 60%-70% with the acellular product (DTPa). However, none of the trials has been large enough to evaluate the rate of rare side effects such as anaphylaxis or encephalopathy in comparison with whole-cell vaccine. What has been established is that HHEs occasionally occur both with acellular pertussis vaccine and with combined diphtheria and tetanus vaccine (without the pertussis component) and that this occurs at a rate of 1 in 10,000 doses compared to about 1 in 1,000 with the whole-cell vaccine. Comparing absolute rates of HHEs between trials (different case definitions) and communities (higher rates of reporting in more versus less advantaged) is difficult, but relative comparisons should be valid. (34)

Reaction	% after 1010 Doses of adsorbed DTP
Local Reactions	
Bruising	52.3
Redness and Swelling	69.8
General Reactions	
Irritability or Vomiting	30.5
Fever	23.4
Persistent Crying	17.2
Persistent Screaming	2.6
Drowsiness	3
Collapse, Pallor, Coldness	0.1
Convulsion	0.1

(Taken from (14) for the Australian Whole Cell Vaccine)

9.2 Serious Acute Neurological Illness

In 1981 a report on the National Childhood Encephalopathy Study (NCES) was submitted to the British Department of Health and Social Security. An abbreviated version of that report appeared in the British Medical Journal during the same year(15).

The study included children aged 0-35 months with serious neurological illnesses. These included encephalitis or encephalopathy, unexplained coma, convulsions lasting longer than 30 minutes or followed by persistent neurological complications, infantile spasms, and Reyes syndrome. The estimated attributable risk of serious neurological disorders occurring within seven days after immunisation with DTP vaccine in previously normal children irrespective of outcome is one in 110,000 injections (95% confidence limits, one in 360,000 to one in 44,000). The corresponding rate for previously normal children with neurological sequelae (related disorder) persistent one year later is one in 310,000 injections (95% confidence limits, one in 5,310,000 to one in 54,000). This is the same as 1/22,000 and 1/62,000 respectively when a course of five DTP injections is assumed. The report said that there were however a number of assumptions made in arriving at these figures. Because of these assumptions and the wide confidence limits, these derived risk figures must be interpreted with caution and cannot be regarded as precise measures.

No significant association was found between serious neurological illness and preceding immunisation with diphtheria and tetanus vaccine.

Research into the neurological impacts of pertussis disease is poor and incomplete. A report was released in 1992 however that reviewed the information compiled by the health authorities in the United States (40). It suggests that 0.9% of children under 12 months of age (1.4% for children under 1 month of age) who are notified as having pertussis will experience some form of encephalopathy.

Information on case definition in this report is almost non-existent and it does not compare results against a control group. So we do not know how many of these cases can be directly attributed to pertussis. In fact previous studies have suggested that a significant number of pertussis related encephalopathy can be related to pre-existing neurological conditions (refer section 5.2). Consequently this information should also be viewed with caution and the figures used as estimate only.

Other information (41) suggests a 50% rate of sequelae in survivors of pertussis related encephalopathy. So in Australia the notification rate of pertussis in children under the age of 12 months is 460/250,000 or 184 in 100,000. Therefore the (current) risk of a child presenting with some form of encephalopathy due to pertussis is 1.7 in 100,000 or 0.85 in 100,000 with sequelae.

If we assume that the current vaccination rate in Australia is 70% and that there are 250,000 children under the age of 12 months:

70% Vaccine Coverage	Total No. of Children with Encephalopathy	No. of Children with Sequelae
Vaccine related	6	3
Disease related	4	2

So, what would happen if we increased the vaccination rate to U.S. levels, say 90%. I have shown the results in the table below and have assumed a flattering 0-12 month incidence reduction from the current 184 in 100,000 to 40 in 100,000 which I have derived from (40).

90% Vaccine Coverage	Total No. of Children with Encephalopathy	No. of Children with Sequelae
Vaccine related	10	5
Disease related	1	0.5

For the sake of completeness I have used the Swedish notification rate to estimate what would occur if the vaccination coverage dropped to 10%. I have shown the results in the table below and have assumed a 0-12 month incidence increase from the current 184 in 100,000 to 630 in 100,000 which I have derived from (7)

10% Vaccine Coverage	Total No. of Children with Encephalopathy	No. of Children with Sequelae
Vaccine related	1	0.5
Disease related	14	7

These comparisons are estimates only and I think the error inflates the disease related encephalopathy numbers quite a lot. Even so I think there is a very slight increase in the risk of pertussis related encephalopathy for unimmunised children.

9.3 Shock and 'Unusual Shock-Like State' (HHE)

In August 1991, a committee of the American Institute of Medicine released a report entitled Adverse Effects of Pertussis and Rubella Vaccines. A summary of the committees key findings appeared in theJournal of the American Medical Association (JAMA) in 1992 (16). For DTP Vaccine they found that there were four events where they could answer yes to the question "can the vaccine cause the adverse event?". They were;

1.   Acute Encephalopathy
2.   Shock and "unusual Shock-Like State (HHE)
3.   Anaphylaxis
4.   Protracted, Inconsolable Crying

The committee attempted to estimate the excess risk associated with immunisation;

9.3.1   Acute Encephalopathy

They used the National Childhood Encephalopathy Study (NCES) referred to in Section 9.2. They somehow came up with a risk of 0.0 to 1.05/100,000 immunisations.

9.3.2   Shock and unusual Shock-Like State (HHE)

HHE stands for Hypotonic-Hyporesponsive Episode and is a descriptive term that applies to an unusual reaction associated with DTP immunization. Characteristically, the infant or child is pale, hypotonic, and unresponsive to his or her parents (17). This is close to the description in Section 9.1 (i.e. Collapse, Pallor, and Coldness) and the risk (i.e. 0.1%) is close to the estimate of the committee of 3.5 to 291 cases per 100,000 immunisations.

9.3.3   Anaphylaxis

The committee estimated a rate of anaphylaxis of two cases per 100,000 injections (or ten per 100,000 children given five doses of DTP). Anaphylaxis is a life-threatening allergic or hypersensitive response to medications and foodstuffs. Reactions that occur almost immediately tend to be the most severe. A doctor injecting vaccines must always have Epinephrine (eg. Adrenaline) close by.

9.3.4   Protracted, Inconsolable Crying

These rates are shown in Section 9.1 for the Australian whole cell vaccines.

My Position (What are the risks to a child's health caused by the pertussis vaccine?):
There are significant known side effects to vaccines and these have been documented for whole cell vaccines and I accept that acellular vaccines present a lower risk of significant side effects.

I think that there could be an argument in favour of the notion that there is a higher risk of neurological damage in the unimmunised however the additional risk would be very small and the evidence is very sparse indeed. This is one area that needs more research.

HOW MUCH IS *NOT* KNOWN ABOUT PERTUSSIS AND THE PERTUSSIS VACCINES?

Research in the following areas is needed to improve our knowledge and understanding of pertussis:

- Dynamics of pertussis transmission, including mother to infant transmission;
- Involvement of adolescents and adults in sustaining endemicity;
- Improved case investigation;
- Serotyping to determine serotype prevalence in the community;
- Monitoring effectiveness of pertussis vaccines in the field, including comparison with new vaccines as they become available and post licensing surveillance;
- Determining the duration of protection for pertussis vaccines, in particular new vaccines and acellular vaccines;
- Local evaluation of new combination vaccines (e.g. those containing Haemophilus influenzae type b (Hib), hepatitis B virus (HBV), diphtheria and tetanus antigens) in terms of immunogenicity, safety and cost-effectiveness;
- A trial of adult formulated pertussis vaccines if initial studies support the view that adults are an important reservoir of Bordetella pertussis infection;

- Development of more appropriate monitoring systems to encompass minor and severe adverse events, incorporating post-marketing surveillance;
- Development of more sensitive, rapid, non-invasive tests for field use;
- Development of serological/biological markers of pertussis immunity and susceptibility;
- Investigation and development of molecular profiling methods (e.g. Pulse Field Gel Electrophoresis);
- Investigation of alternative antibiotic therapy, particularly other macrolides and once daily or single dose therapy;
- Greater input into education through promotional campaigns;
- Improved monitoring and evaluation of parent perceptions, risk/benefits of immunisation and educational or promotional activities. (33)
- The efficacy of homeopathic prophylactic remedies.

10.1   Sudden Infant Death Syndrome (SIDS)

The next 9 paragraphs have been taken from an article that appeared in the "American Family Physician" (18).

SIDS is defined as the sudden death of an infant under one year of age that remains unexplained after thorough case investigation, including performance of a complete autopsy; examination of the death scene; and review of the clinical history.

After decades as the number one cause of death in infants between one week and one year of age, SIDS is declining in the United States. The rate has declined from an average 1.3 to 1.4 SIDS deaths per 1,000 live births to about 0.85 SIDS deaths per 1,000 live births (in Australia the rate is approximately 0.55 per 1,000 live births, refer section 13 below). This change was not brought about by breakthrough discoveries of the cause or causes of SIDS, but by public education about the risk factors for SIDS.

The major risk-reduction measures supported by available scientific research are

1.   having healthy (my highlight) babies sleep in the supine position (on the back). Studies suggest that the prone sleeping position may increase SIDS risk by increasing the probability that the baby rebreathes his or her own expired gas, leading to carbon dioxide build-up and low oxygen levels; by causing upper airway obstruction; by interfering with body heat dissipation, leading to overheating; and by a variety of other proposed mechanisms.

2.   not exposing babies to cigarette smoke, either during pregnancy or after birth. Smoking during pregnancy exposes the developing foetus to toxins and other potentially harmful effects of cigarette smoke. In addition, increasing evidence suggests that exposing the baby after birth to cigarette smoke also increases the risk of SIDS. The increase in SIDS risk appears to be related to the "dose" of passive smoke exposure - the greater the exposure to smoke both before and after the birth, the higher the risk of SIDS.

3.   making the sleeping environment as safe as possible. Studies going back three decades have indicated that soft bedding material may be hazardous for young infants. Recommendations for cot safety include making sure the cot is in good working order with no missing or broken parts; being sure that the mattress is in good condition and is the proper size for the crib; and positioning the cot so that curtain or blind cords do not hang down into or near the cot and pose a strangulation hazard.

Because of the potential hazards for overheating, it is generally recommended that for sleep, a baby should be lightly dressed and covered with a sheet or thin blanket, and the room temperature should be such that it would be comfortable for an adult in a short sleeved shirt. Swaddling and tight "tucking in" of infants is not recommended.

4. Breastfeeding. After numerous studies on breast-feeding and SIDS, results are conflicting, and the relationship is still unclear. What is known is that breastfeeding offers many potential physiological and psychological benefits to mother and infant and is, in general, preferable to bottle feeding and recommended.

The important thing to realise out of all of this is that SIDS is unique because, by definition its major presenting symptom is unexplained death, the diagnosis is based entirely on what is not found. SIDS is, in other words, a diagnosis of exclusion.

This diagnosis of exclusion is an important point. At best SIDS is caused by a number of risk factors of which the 4 points, above, seem to be the most important. Another study on SIDS (or "cot death") investigated unexplained deaths in Sheffield between June 1972 and March 1976. The results were reported in the Lancet in August 1979 (19).

The report showed quite clearly that the risk of SIDS increased dramatically when the child presented with certain symptoms. The top nine symptoms (in order of importance) in the report were;

1. Snuffles
2. Cough
3. Irritable
4. Vomiting
5. Diarrhoea
6. Sleepy
7. Rash
8. Change of Crying
9. Fever

The risks increased further when the child was premature, a twin or had a younger sibling who had died of SIDS.

There are disturbing similarities between this table and the one shown in Section 9.1.

For an even more alarming and well-researched link between SIDS and the DTP vaccine I refer you to Viera Scheibners book (20). Her work revolves around non-specific stress syndrome and its relationship to SIDS. Her research shows quite clearly that the danger period for a child's response to a vaccine is not 1,2 or even 7 days after injection it is 2 or 3 weeks after the shot. Scheibner's tabulations show a marked increase in symptoms (shown above) in children 16 to 17 days after a DTP shot. She then correlates this finding with raw data from peer-reviewed articles on SIDS victims. Her methods seem quite sound and reasonable.

Can I put a number on the rate of SIDS cases caused by DTP vaccines? The answer is no because SIDS has a definition of exclusion (i.e. no one knows what caused it). Can anybody say that DTP vaccines do not cause SIDS? No, for the same reason.

There is also evidence that DTP related deaths have been classified as SIDS in the past. In March, 1979 the U.S. Surgeon General intervened and a manufacturer of DTP vaccines withdrew all unused doses of a batch number 64201 after 8 children died (classified as SIDS) within 7 days of their first dose from this batch of DTP vaccine(21).

I have read a number of studies that dispel the link between SIDS and DTP vaccines and the raw data in these reports seem to support Viera Scheibner's findings (i.e. clustering of SIDS deaths 16 days after DTP vaccination). I have also read the report that suggests a "protective effect of DTP immunisation" based on a less than expected occurrence of SIDS cases 0-7 days after the DTP shot. Their "most plausible explanation for the decreased rate of SIDS in the period immediately after immunization is that children may be immunised when they are in better health and that this healthier state is associated with a lower risk of SIDS."(22)

I do not think that there is a direct correlation between vaccine uptake and SIDS. I do know that **IF** on average two or three of the 139 SIDS cases each year is caused by a pertussis vaccine then, referring to the death rates due to pertussis shown in section 8.2, more children are killed by the vaccine than are saved by it!

10.2   DTP Vaccines and Asthma

In August 1994 Michel Odent managed to get a letter printed in JAMA (23). In the letter was a review of a study examining the criteria of health in a group of children and adolescents. I have quoted a portion below;

"Among the 243 immunised children (mean age, 8.12 years), 26 were diagnosed as having asthma (10.69%), compared with four (1.97%) of the 203 children (mean age, 7.59 years) who had not been immunised. The relative risk is 5.43 (95% confidence interval, 1.93 to 15.30). The significance is at the $P=.0005$ level."(23) These results are statistically significant ($P<0.05$). According to these results at best you are only twice as likely to get asthma. At worst you could be 15 times more likely.

I decided to take a visit to one of the many web sites available on the subject. The first one I tried is run by the National Asthma Campaign (NAC) at http://hna.ffh.vic.gov.au/asthma/ and that is where I compiled most of the information in the two paragraphs that follow (Note: this web site may have changed).

The basic facts are that 1 in 4 children contract Asthma. It is among the 10 most common reasons for seeing a GP. Each year about 750 people die of Asthma, about 3 of these are 0-4 years of age. Asthma is a disorder of increased sensitivity of the airways. The sensitive airways react when exposed to certain triggers. They constrict, become inflamed and excess mucus is produced. Symptoms of the attack include cough, wheeze (whistling sound in the chest), chest tightness and shortness of breath.

The graph (see below) shows the death rate due to Asthma (deaths/100,000) from 1920-1995. In order for this to be relevant to a discussion on pertussis, it needs to be compared to the pertussis vaccine coverage rate and the history of Asthma prevention initiatives in Australia. Data on vaccine coverage can only be inferred by history before 1989 when the Australian Bureau of Statistics started recording vaccination data. Even the ABS information after this date is questionable due to different methodologies used in assessing coverage and the absence of a nationally agreed definition of full immunisation coverage.

What follows is a short history of key developments in vaccine coverage and Asthma control:

> In Australia the first pertussis vaccine was manufactured by the Commonwealth Serum Laboratories (CSL) in the 1920s and a more effective vaccine was released in 1953. In 1954, the National Health and Medical Research Council (NHMRC) recommended immunisation of infants using the new triple antigen vaccine.

Global concerns regarding the safety of pertussis vaccines in the early 1970's <u>lead to a reduction in vaccine coverage</u> and culminated in the 4th dose of DTP being dropped from the schedule in 1978, <u>leading to a probable increase in coverage.</u> Australia reintroduced the 18-month booster in 1986 which <u>may have lead to another decline in coverage.</u>(33).

In 1990 the National Asthma Campaign started a comprehensive program of education regarding Asthma and ways of mitigating the harmful effects of the disease. The success this program has had in reducing the Asthma death rate is worthy of great praise.

A 5th dose of pertussis vaccine was introduced in Australia in 1994, on the basis of evidence of increased pertussis incidence (particularly in school aged children) (33).

The way the graph follows the history of pertussis vaccine in Australia is striking and adds further credence to the research by Michel Odent. As with SIDS **IF** on average two or three of the 400 – 750 Asthma deaths each year is caused by a pertussis vaccine then, again referring to section 8.2, more children are killed by the vaccine than are saved by it! Also, **IF** there is a direct link between asthma and the pertussis vaccine then it is hard to know what is worse; a bout of pertussis or a potentially life long struggle with asthma.

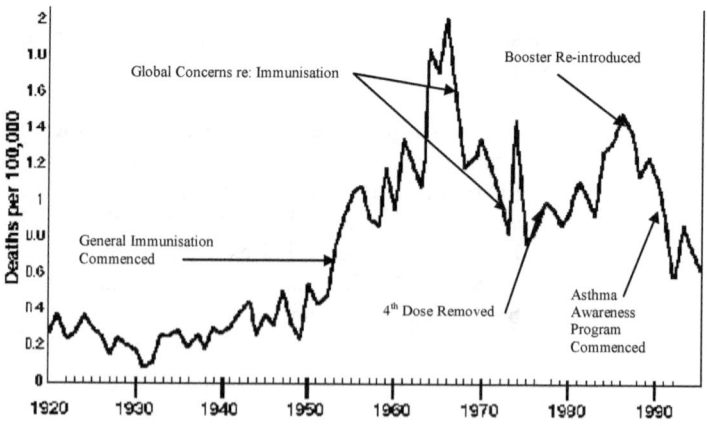

10.3   Invasive Bacterial Infections

In 1986 some Swedish researchers assembled a trial using 3801 children aged 6 to 11 months to test the efficacy of acellular pertussis vaccines. All the children in the trial received a first dose of vaccine or placebo between February and March of 1986 and most received a second dose 7 to 13 weeks later. The trial participants were followed up 7 to 9 months later. Four vaccinated children died during the 7 to 9 months follow-up as a result of invasive bacterial infections. Quoting from the report;

"Mortality from invasive bacterial infections was estimated to be 1.1 deaths/1000 vaccinated children compared with 0.09/1000 (5 of 57 000) among remaining nonvaccinated eligible children"(24)

These risks estimates are based on the particular study and come with no indication of statistical significance or confidence ranges. The deaths may have been caused by a single batch of the Japanese acellular vaccine used in the trial. There are however other indicators that the DTP vaccine may increase the incidence of invasive bacterial infections.

A report in the New England Journal of Medicine in 1971 showed a steady, unexplained increase in the incidence of Haemophilus influenzae meningitis at a children's hospital in Pittsburgh. The increase spanned 3 decades commencing around the time DTP vaccines were introduced.(25) Another study in Gloucestershire in 1986 showed an increase in the occurrence of Meningococcol disease in the 1980's. The figures for England and Wales in the same report seem to correlate well with the fluctuating uptake of the DTP vaccine during the same period. (26)(8).

10.4   How unhealthy are the Vaccines?

I'll not dwell too much on this issue mainly because I think a discussion on the specific constituents of vaccines is of little value. No matter which way you look at it each injection introduces a man made substance into a young body. There is no clear understanding of what longer term impacts this has on the immune system.

As an example there are two ingredients in a whole cell DTP vaccine that have caused debate in the past;

- Thiomersal is a toxic mercury compound. Mercury, among other things, suppresses the immune system, increasing susceptibility to infections. As I understand it all vaccines on the current Australian Vaccination Schedule for children under five are now thiomersal-free.

- Aluminium Phosphate is listed in the EPA's Extremely Hazardous Substances List, corrosive to the eyes, skin and mucous membranes.

While these materials are worth a mention, I believe the main concern with vaccines relates to the quality of the manufacturing and distribution process. I work in a manufacturing environment and I know that no process is perfect. How often does the vaccine manufacturing process move outside of prescribed tolerances? When this happens how much of this substandard material is released as "good stock"? If this happens what is the impact of this substandard material if it is injected into a child's bloodstream?

There is already evidence that this has happened and that this has resulted in death (refer to Section 10.1)

My Position (What are the risks to a child's health caused by the pertussis vaccine?):
I believe that we do not know enough about the long term effects of vaccines on our children. In fact I think there is enough evidence to suggest that vaccines may be weakening the immune system and exposing our children to greater risks associated with SIDS, Asthma, and bacterial infections.

The only longer term study (not intended as a review of the pertussis vaccine) has revealed alarming relationships between pertussis vaccines and asthma. What would be the outcome of a more focussed study comparing key health indicators for immunised and unimmunised children and why hasn't this fundamental study been carried out?

I also think that many serious adverse reactions to vaccines go unreported. Which doctor will advocate vaccines to their patient and then 2 weeks later undermine his position completely by admitting that a reaction such as SIDS or a neurological reaction could have been associated with the vaccine.

WHAT WOULD HAPPEN TO VACCINE COVERAGE AND INCIDENCE OF THE DISEASE IF PEOPLE WERE GIVEN INFORMED CHOICE?

I think a pro-choice environment would have an immunisation acceptance rate of about 60%-70%. This is a guess based on the reading I have done to date. There is however the England/Wales experience to draw from where, in the mid seventies, immunisation acceptance dropped to reach its lowest level of 30% in 1978 (8) after adverse publicity regarding side effects to the Whooping Cough vaccine. Hence I think less hysteria in a pro-choice environment is worth another 30%-40%.

Australia currently has a vaccine coverage rate in the vicinity of 60 – 80% (estimated using (32) although accurate figures are hard to find because of problems with definition). So I suspect that reverting to an environment of responsible choice will not increase notification rates significantly.

My Position (What would happen to vaccine coverage and incidence of the disease if people were given informed choice?):

I think a pro-choice environment would cause a vaccine coverage rate of between 60% and 70% that is not significantly different from the current estimated immunisation rate. I believe that reverting to an environment of responsible choice will not increase notification or incidence rates significantly in Australia.

HOW IS PERTUSSIS TREATED?

In this section I have covered the treatment of both Pertussis and Croup and used (38), as my main reference. I have included croup because it is such a similar condition to Pertussis and because it can be very worrying for someone who has not vaccinated their child.

12.1 Pertussis Treatment

1. Interventions

Routine treatment consists of bed rest, adequate nutrition, and adequate amounts of fluid. Erythromycin may be prescribed to reduce transmission or to control secondary infection (refer Section 14.2). Hospitalisation may be necessary for infants and children with severe or prolonged paroxysms and for those with dehydration or other complications. Oxygen may be needed to relieve dyspnea and cyanosis; intravenous therapy may be necessary when prolonged vomiting interferes with adequate nutrition. Intubation is rarely necessary but may be lifesaving in infants if the thick mucus cannot be easily suctioned from the air passages.

The pathophysiology of the central nervous system involvement is not clear. There are probably multiple factors including hypoxemia, bleeding, hypoglycemia, and direct neurotoxic effects of pertussis toxin. Attention to preventing hypoxemia and hypoglysemia is essential (41).

2.   Nursing Considerations

Severe paroxysms in an infant may require oxygen, suction, and intubation. The child needs to be kept calm and protected from respiratory irritants such as dirt, smoke, or dust. Over-stimulation, noise, or excitement may precipitate paroxysms. Adequate nutrition and adequate fluids are encouraged through frequent small feedings.

Central nervous system involvement (refer Section 9.2) includes apnea, seizures, and encephalopathy. Apnea is especially common in young infants. Seizures occur alone or as part of an encephalopathy that sometimes progresses to coma. Supportive care and anticonvulsant medications should be instituted. (41)

12.2   Croup Treatment

Children endure many upper respiratory tract infections during their lifetime. Normally these infections are short lived and don't cause much alarm. They are usually due to viruses and cannot be cured. Treatment is matter of reducing the symptoms with things such as Panadol for temperatures, and decongestants (used wisely) to make life more comfortable.

But sometimes the infection can be more serious, putting the child at risk. One such infection is Croup, which is stressful for both the child and the parent. It can be more stressful for the parent that has decided not to vaccinate the child against pertussis. This is because the symptoms of this disease can be very similar to pertussis. Croup is a viral infection affecting the larynx (voice box) and trachea (upper part of the windpipe). Croup is most common in children between the ages of six months and three years. It is worse at night.

The affected child has a dry 'barking' cough and may develop stridor – a high-pitched wheeze or grunt – when it breathes. In small children the upper airway can become obstructed. This is a life-threatening complication requiring urgent hospital admission.

1.   Interventions

The treatment of croup in its mild stages consists of reassuring the child – a frightened child takes rapid shallow breaths which don't get much air into its lungs - bed rest, adequate fluid intake, and alleviation of airway obstruction to ensure adequate respiratory exchange. Children with mild infections are usually managed at home with supportive measures, such as use of vaporizers, humidifiers, or steam from hot running water in an enclosed bathroom to reduce the spasm of the laryngeal muscles and to free secretions. Hospitalisation is indicated for children with high temperature; progressive stridor and respiratory distress; and hypoxia, cyanosis, or pallor. Endotracheal intubation and tracheostomy may be necessary. Humidity and oxygen are usually prescribed. The vital signs are continuously monitored; changes in pulse and respiration may be early signs of hypoxia and impending airway obstruction. Fluids are often given intravenously to reduce physical exertion and the possibility of vomiting, with its attendant increased risk of aspiration. Corticostaroids and inhaled racemic epinephrine are often used. Other drugs, such as expectorants, bronchodilators, and antihistamines, are rarely used, and sedatives are contraindicated, because they exert a depressant effect on the respiratory tract.

2.   Nursing Considerations

The primary focuses of nursing care are to ease breathing by providing humidity and to monitor continuously for signs of respiratory distress and impending airway obstruction, with intubation and tracheostomy equipment kept readily available. To conserve the childs energy and to reduce apprehension, the nurse encourages rest, disturbs the child as little as possible, remains in constant attendance, provides comfort with a familiar toy or other device, and encourages parental involvement whenever possible. Fever is usually reduced by the cool atmosphere of the mist tent; antipyretics are given as needed. To prevent chilling, frequent changes of clothing and bed linen are often necessary in the humid environment. In most children the condition is relatively mild and runs its course in 3 to 7 days. The infection may spread to other areas of the respiratory tract and may cause complications such as bronchiolitis, pneumonia, otitis media. The most serious complication is laryngeal obstruction, which may cause death.

My Position (How is Pertussis treated?):
The treatment of pertussis consists of bed rest, adequate nutrition, and adequate amounts of fluid. Erythromycin may be prescribed to reduce transmission or to control secondary infection. Hospitalisation may be necessary for infants and children with severe or prolonged paroxysms and for those with dehydration or other complications. Oxygen may be needed to relieve dyspnea and cyanosis; intravenous therapy may be necessary when prolonged vomiting interferes with adequate nutrition. Intubation is rarely necessary but may be lifesaving in infants if the thick mucus cannot be easily suctioned from the air passages.

The child needs to be kept calm and protected from respiratory irritants such as dirt, smoke, or dust. Overstimulation, noise, or excitement may precipitate paroxysms. Adequate nutrition and adequate fluids are encouraged through frequent small feedings.

HOW DOES THE RISK OF DYING FROM PERTUSSIS COMPARE TO OTHER RISKS?

There were approximately 250,891 births registered in Australia in 1998 of which 1336 were still born. The table below shows how the remaining children will eventually die (given the current environment). This information has been extrapolated from (39).

Cause of Death	Percentage	Number of Persons
Malignant neoplasms (cancer)	27.2	67879
Ischaemic heart disease	21.9	54653
Cerebrovascular disease (stroke)	9.4	23458
Chronic obstructive pulmonary disease and allied conditions (including asthma, emphysema and bronchitis)	4.8	11979
Accidents (inc. motor vehicle)	3.8	9483
Pneumonia and influenza	3.6	8984
Diabetes mellitus	2.2	5490
Diseases of arteries, arterioles and capillaries	2.2	5490
Suicide	2.1	5241
Hereditary and degenerative diseases of the central nervous system	1.6	3993
All Other Causes	21.2	52905
Total:	**100**	**249555**

Also in 1998 1,252 children died before they reached their first birthday

Cause of Death (0-12 months)	Percentage	Number of Persons
Congenital anomalies.	27	338
Certain conditions originating in the perinatal period (i.e. any time within 28 days of birth)	45.6	571
Sudden death, cause unknown (SIDS)	11.1	139
All Other Causes	16.3	204
Total:	100	1252

In 1998 there were no deaths caused by pertussis (although the average is 1.8 per year).

My Position (How does the risk of dying from pertussis compare to other risks?):
I do not believe that the risk of a child dying from pertussis is reduced significantly by vaccination. I also believe that issues such as SIDS are impacted by the health of a child's immune system. If a child's immune system is compromised in any way during the crucial first 12 months then I think that this does not bode well for its short and long term health prospects.

65

WHAT OTHER WAYS CAN YOU MINIMISE THE RISKS OF YOUR CHILD CONTRACTING PERTUSSIS AND TRANSMITTING THE DISEASE TO OTHER CHILDREN?

14.1   Be aware of cases in community

Pertussis is usually well publicized within a community mostly due to the surprise of most that they or their children can still catch the disease even though they are vaccinated. Usually pertussis hits the community as a mini outbreak until the public health authority can act, trace patients and then control the outbreak. If you hear of a series of cases within your community then be aware of the precautionary measures described below. Information in sections 14.2 and 14.3 is taken from (33). Again, I must stress that I am not a medical practitioner and you should seek professional advice if you think your child is a significant contact.

14.2 What to do if you or your child becomes a pertussis contact?

If you or your child become a significant contact as broadly defined in section 3 then you should seek antibiotic prophylaxis. The purpose of antibiotic prophylaxis is to prevent disease in those exposed to a case and to decrease transmission to those at highest risk of severe disease namely those under 12 months of age.

While vaccination against pertussis is effective in reducing the incidence of the disease, vaccination prevents NEITHER colonization NOR transmission of B. pertussis. This means that those who have been vaccinated can still be infected and transmit B. pertussis. Erythromycin therapy for cases and prophylactic treatment of contacts may therefore be beneficial. Although there have been no controlled, blinded studies of the efficacy of erythromycin in preventing transmission of pertussis to contacts, a number of retrospective studies suggest that erythromycin given to contacts within two weeks of onset of cough in the index case (the case to whom the contact was exposed) halts transmission and prevents disease in contacts.

The public health authority should be advised of all cases of pertussis and will generally approach prophylaxis in the following manner:

> It is the public health authorities' role to provide advice for all contacts and caregivers in high-risk settings such as family day-care and child-care regarding appropriate prophylaxis. In other settings, prophylaxis should be undertaken after an assessment of risk (Section 3). Contacts under one year of age and children not up-to-date with the recommended vaccinations against pertussis should routinely receive prophylaxis, as should infants less than one year of age who have significant contact in other community settings.

> Erythromycin (40 to 50mg/kg/day in four divided doses orally, maximum 1g/day for 10 days) should be

recommended for all household contacts and other contacts in high-risk settings where there are susceptible individuals, especially infants. Erythromycin should be given as soon as possible, and no later then 14 days, after the recipients first contact with a primary case during the infectious period (in high risk exposure settings prophylaxis may be considered for up to 21 days from a recipients first contact with a primary case) (American Academy of Pediatrics, 1991). Where infants are at risk within a household or other high risk setting antibiotic treatment of contacts should NOT be delayed pending confirmation of the diagnosis. Note that the infectious period for untreated cases is up to 21 days from onset of cough while for cases receiving antibiotic treatment the infectious period is five days after the commencement of Erythromycin therapy.

Maternal antibodies do not protect newborns against pertussis. Management of pregnant contacts must be on an individual basis and should be discussed with local experts. Erythromycin is poorly tolerated during pregnancy because of its gastrointestinal side effects. However, infants of mothers who become symptomatic with pertussis up to three weeks before labour have an extremely high risk of the disease. Household contacts that are pregnant should therefore be offered prophylaxis. If treatment is not tolerated or not complete by the time of delivery, both mother and baby should be given prophylaxis after delivery. In such cases, newborn babies should be given erythromycin syrup 40mg/kg/day in four divided doses for 10 days from birth.

Erythromycin must be considered for each new episode of close exposure unless the contact was receiving erythromycin at the time of exposure.

The earlier the Erythromycin therapy is started, the greater the chance that the therapy will alter the disease. If erythromycin is begun during the prodromal/incubation phase, the disease is aborted. If the therapy is begun in the catarrhal or early paroxysmal phase, there is amelioration of symptoms. Therapy later in the paroxysmal phase is unlikely to alter the disease, but is still recommended to clear the organism from the airway and thus render the child noncontagious.(41)

Vaccination of contacts will not prevent infection in those who have already been exposed to the disease and are not an alternative to antibiotic prophylaxis. Also, according to the health department, vaccination is not usually indicated for children eight years of age or older.

14.3   Exclusion of contacts

The infectious period for untreated cases is up to 21 days from onset of cough while for cases receiving antibiotic treatment, the infectious period is 5 days.

Cases or pertussis should be excluded from school, preschool, day-care or other settings where there are susceptible individuals, especially young children and infants, for 21 days from the onset of illness or until they have received at least five days of a 10 day course of erythromycin.

Unimmunised household contacts less than seven years of age should be excluded from school, preschool, day-care or other settings where there are susceptible individuals, especially young children and infants, for 14 days after their last exposure to infection or until they have received at least five days of a 10 day course of erythromycin. Close child-care contacts (broadly defined in section 3) should also be excluded for 14-days from their last exposure to infection or until they have received at least five days of a 10 day course of erythromycin.

My Position (What other ways can you minimise the risks of your child contracting pertussis and transmitting the disease to other children?):

While the risk of an unimmunised child contracting pertussis can never be eradicated, it can be reduced. By being aware of 4 yearly cycles and the seasonality of the disease we can be vigilant for signs of local outbreaks. If your child is significant contact then antibiotic prophylaxis may be employed to reduce the infectious period and to reduce the symptoms to the extent that the outward signs of the disease can be avoided all together if administered early enough.

HOW IS MONEY INVOLVED IN THE PROMOTION OF THE PERTUSSIS VACCINE?

The following information is taken from (34).

Two processes contribute to making drugs and vaccines accessible in Australia:

i.   the marketing approval (licensing) process, through the Therapeutic Goods Administration (TGA), which considers the quality, safety and efficacy of pharmaceutical products; and

ii. the process for subsidizing the cost of drugs through inclusion on the Pharmaceutical Benefits scheme (PBS), for which the data required are comparative efficacy and comparative cost-effectiveness.

The PBS was established in 1953 and has been a remarkably robust political policy, the aim of which is to provide access to essential drugs. Drugs are evaluated for listing on the PBS by the Pharmaceutical Benefits Advisory Committee (PBAC), which is a powerful advisory committee, the minister cannot make a decision to list a drug unless the PBAC has recommended that (s)he do so.

15.1 Requirement for the PBAC to consider comparative cost-effectiveness.

An amendment to the National Health Act in 1989 established the requirement for the PBAC to consider comparative cost-effectiveness in making recommendations to the Minister. The PBAC guidelines for comparative cost-effectiveness, first developed in 1990-1991, are now in their second edition and consist of two parts:

    i.   establishing the relative clinical benefit of any new product, and;

    ii.   evaluating that benefit.

15.2 Cost-effectiveness evaluation for vaccines.

Vaccines have been required to be approved for marketing through the TGA, but have not generally been subject to evaluation of comparative efficacy and cost-effectiveness, either because they were PBS listed for individual use prior to the introduction of current guidelines or because funding for population use (as for NHMRC schedule vaccines) has been provided under separate processes. The current acellular vaccine (Infranrix) was the first vaccine subject to an economic evaluation and presented a number of new issues to the PBAC.

Although vaccines are used for prophylaxis rather than treatment, they are not alone in that, drugs for osteoporosis and hypertension, for example, are also prophylactic. Probably of more difficulty for evaluating vaccines is the question of community as well as individual benefit, which is not usually part of a drug evaluation.

15.3 Economic evaluation of Infanrix versus whole-cell vaccine

Ms Michelle Burke, health economist with SmithKline Beecham (SKB), led the team that conducted the economic analysis of the vaccine (Infanrix) which was submitted to the PBAC. She presented the methodology and summary findings of the economic analysis, but was unable to present detailed data because of commercial confidentiality issues.

The team working on the analysis developed a model with several key assumptions:

i.   the efficacy of Infanrix (DTPa) and the CSL whole-cell vaccine (DTPw) was equivalent *(questionable refer section 6)*;

ii. the better tolerability of Infanrix would result in improved coverage rates *(questionable – I think the public awareness of the difference is not significant)*; and

iii. increased coverage would lead to fewer cases and deaths from pertussis *(questionable refer section 5.2 and section 8)*.

Apparently, the model developed was complex. It included changes over time in both the probability of infection, to account for cyclical epidemics, and coverage rates. It also included consideration of children of differing ages and immunisation histories. No empirical data were available for a number of variables in the model (for example, improvement in coverage from use of Infanrix) and values for these variables were derived from the consensus opinion of an expert panel. Sensitivity analysis was used to examine the changes that occurred in the model estimates when different values, within the plausible range of values, were substituted for the value selected as baseline for the model.

The model estimated that the cost per pertussis infection prevented was less than $3,000, and the cost per life year gained was less than $25,000. These estimates were sensitive to changes in the following three factors: baseline coverage rates, coverage with Infanrix, and the probability of pertussis infection. Where less favourable estimates were obtained with sensitivity analysis, estimated costs did not increase to unacceptable levels.

The estimates for Infanrix (under $3,000 per infection averted, and under $25,000 per life year gained) *where* considered in the context of previous decisions about other drugs. A league table of estimated cost per quality adjusted life year (QALY) for various drugs presented to the PBS since 1990 suggests that estimates of $20,000-$30,000 per QALY are acceptable and estimates of more than $100,000 are unacceptable. The estimates for infanrix were well within the boundaries considered by PBAC when evaluating drugs.

As submissions on cost-effectiveness for the PBAC are protected under secrecy provisions of the National Health Act, more detail regarding models used in the cost-effectiveness of vaccines are unavailable.

15.4   How does SmithKline Beecham benefit?

I am not sure how much each injection of DTP is worth however, a similar vaccine in the USA costs about $30 per shot. If there are 250,000 live births each year and assuming that acellular vaccines will eventually replace all five DTP shots on the Australian immunisation schedule then this is a potential business to SmithKline Beecham that could eventually exceed $37 million per year.

Also, if you look carefully at the health pamphlets available at your local doctors' surgery or chemist you will find immunisation information booklets generously sponsored by SmithKline Beecham. The same pamphlets usually espouse the virtues of another SmithKline Beecham product called Panadol in alleviating any of the side effects associated with vaccines. The additional Panadol sales to SmithKline Beecham would also be substantial.

My Position (How is money involved in the promotion of the pertussis vaccine?):
Pertussis vaccines and other related pharmaceutical products are big business, currently dominated in Australia by CSL and SmithKline Beecham.

The PBAC considers comparative cost-effectiveness and makes recommendations to the Minister based on assumptions and empirical models developed by the potential vaccine supplier. I think this is a conflict of interests.

These studies are then protected by commercial confidentiality arrangements. So we (the general public) are unable to gain access to the analysis that justifies vaccinating our children.

WHAT IS THE ROLE OF THE GOVERNMENT AND VACCINE AWARENESS ORGANISATIONS IN PROVIDING INFORMATION?

16.1   Government

The main vaccine information tool for the health department is The Australian Immunisation Handbook published by the National Health and Medical Research Council (35). The handbook focuses on the mechanics of vaccination and also presents some facts on the risks associated with both the diseases and their vaccines.

In my view the handbook is one sided and does not present an objective view of vaccination. It does not give context for the reader and is very confusing. It provides no references for further reading and statements are not referred appropriately to the bibliography. I have all the references obviously used by the handbook (in relationship to pertussis) however the bibliography only refers to a handful and even then only the references that support the government pro-immunization view.

I have quoted a couple of references from the handbook that relate to the risks associated with pertussis and the pertussis vaccine with commentary on how (to me) they are misleading. I could quote the entire section and make some comment on how misleading I think each sentence is although I have selected just two examples.

Quotation One

"About 1 in 200 whooping cough patients under the age of 6 months dies from pneumonia or brain damage"

I believe that a lot of the references in the handbook are misleading mainly because a lot of important information is missing. First of all the reader is given no context in which to make an assessment of this comment. The first obvious question is how likely is my child to get pertussis in the first 6 months? On average about 460 children under 12 months is notified to the health department as having whooping cough. So if there are about 250,000 live births per year then on average your child has a 0.184% chance of getting pertussis before he or she reaches 12 months. Therefore the chance of that child getting pertussis and dying before the age of 12 months is 0.184% X 0.5% =0.00092% or less than 1 in 100,000. The odds are even lower for a child under the age of 6 months.

Quotation Two

"The overall mortality from pertussis is 0.3% but the mortality in babies under 6 months of age is higher (0.5%)."

At best this is misleading and at worst it is wrong. I think that parents look at this quotation and expect that 0.3% of the population die from pertussis.

Between 1993 and 1998 no one over the age of 12 months died from pertussis or complications from the pertussis disease. During this period the average number of notifications per year was 5848 (460 were under 12 months) and the average number of hospitalisations was 961 per year (460 were under 12 months).

On average 1.8 children under 12 months died so the 0.5% mortality rate for children under 6 months is credible ONLY AS A PERCENTAGE OF THE NOTIFICATION (1.8 divided by 460) OR HOSPITALISATION (1.8 divided by 460) RATE.

16.2   Vaccine Awareness Organisations

Vaccine awareness organisations in their purest form are worthwhile. Their mandate is one of providing unbiased vaccine information to the community. If the government performed its role appropriately these organisations would be unnecessary.

Vaccine awareness organisations also struggle to provide information in the appropriate context. Books and articles provided will often be very specific and the information is generally harder to combine into a bigger picture. Also, the information from vaccine awareness organisations tends towards an anti-vaccination view and, while in some respects this is probably necessary to offset the pro-vaccination stance of the government, this can often work against their credibility.

On the whole the vaccine awareness organizations do a great job considering that they obviously attract no government funding.

My Position (What is the role of the government and vaccine awareness organisations in providing information?)

If the government was doing its job correctly there would be no need for vaccine awareness groups.

MY FINAL POSITION

Pertussis is a long disease that can be distressing for the patient and carer particularly during the paroxysmal phase. Laboratory confirmation of the disease is best achieved via a nasopharyngeal swab that can also be distressing for a child. Epidemiological confirmation of existing symptoms can be enough to commence treatment.

Bordetella pertussis is highly infectious. It may be spread from person to person by close contact usually by respiratory aerosols. Identifying a significant contact must be on an individual basis and take into consideration the degree of risk to the individual and the specifics of the exposure.

The incidence of pertussis in Australia is anywhere between 35 and 350/100,000 depending on the way the incidence is assessed. The incidence increases up to threefold in the summer months and yearly rates double every 4 years.

Rates of notification and hospitalisation are much higher (as high as 2% for children under 1 year of age) for infants and young children.

In it's worst form pertussis can be a frightening disease, particularly for the young. The severity of a case of whooping cough can be expressed in 3 ways - that is, the duration of the illness, the maximum number of coughing spasms per 24hours, and the incidence of complications.

The disease usually lasts 50 days travelling through different phases of severity. At the peak of his/her illness a child will experience about 12 coughing spasms in a 24-hour period. About 16% of the people notified as contracting whooping cough in Australia will be admitted to hospital. Of those admitted to hospital about 70-90% are under 5 years, 50% are aged less than 12 months and 15% less than 3 months which reflects the increased severity of the disease in the very young.

About 20% of hospital admissions will get pneumonia and perhaps 0.5% will experience generalized fitting. 7% of the hospital admissions will be admitted to intensive care, 4% will require artificial ventilation (13) (of those requiring ventilation over 90% will be under 6 months of age).

0.3-0.5% of hospital admissions in Australia due to whooping cough end in death. While nearly all patients that die from Pertussis will have contracted pneumonia, death will occur from respiratory failure, usually from apnoea rather than from secondary infection or aspiration. Of those that die from complications relating to whooping cough, nearly all will be younger than 9 months. The evidence suggests that those older than 9 months will have other longer term health issues. It seems that prematurity may be a risk factor for pertussis-related pneumonia and death.

The evidence also suggests that vaccination reduces the main severity indicators (i.e. disease duration, coughing spasms and complications) by 10-30%.

The effectiveness (efficacy) of vaccines can vary anywhere between 35% and 97% depending on the type of vaccine and which of the 3 serotypes of the organism are prevalent at the time (type 1,2,3; type 1,2; and type 1,3).

Acellular vaccines are proving to be more effective than whole cell vaccines. There is no specific data on the 3-component acellular vaccine used in Australia however there is enough evidence to suggest that it's efficacy in the first couple of years after the final shot would lie between 70% and 97%.

In the absence of other data it can be assumed that the protective effect of acellular vaccines over time is similar to whole cell vaccines. In which case the protective effect will have reduced to 0% by the time a child reaches his/her mid to late teens.

Also, I could find no information to support the notion that the protective effect of vaccines is anything other than 0% before the final shot is administered at the age of 6 months. My belief is that this is probably too conservative.

The pertussis vaccine will never provide herd immunity because it is not 100% effective and its protective effect reduces significantly with time. Children under 1 year of age are the most at risk should they contract Pertussis however, the protective effect of the vaccine is considered 0 until all shots are completed at 6 months.

Consequently, the primary goal of the pertussis vaccine is to reduce exposure of young children (particularly those under 1 year of age) to the disease. Then if exposed, the vaccine offers the children (older than 6 months) one last barrier in terms of protection against infection.

The benefits of immunising siblings within a family in order to reduce the exposure risk to a child under one year old (therefore minimising the pertussis related death rate) is questionable considering the child's high risk exposure to it's mother and father who are essentially unimmunised. More research is required in this area.

There will always be a pool of unimmunised adults and adolescents regardless of the vaccine coverage unless the longer term protective effect of acellular vaccines is any better than whole cell vaccines. If the acellular vaccines perform similarly to whole cell vaccines then as the coverage rates increase the attack rates for immunised children will not change while the attack rate for unimmunised children will probably increase slightly as the disease becomes more virulent.

The overall notification rates will reduce in a linear fashion as vaccine coverage increases commensurate with the effectiveness of the vaccine, the prevalent serotype and demographics.

Also considerations such as climate, economic/living conditions, public health budgets, type of disease, case definitions and other yet unknown factors will all have an effect on the absolute minimum notification rate achievable within a given country.

There are significant known side effects to vaccines and these have been documented for whole cell vaccines and I accept that acellular vaccines present a lower risk of significant side effects.

I think that there could be an argument in favour of the notion that there is a higher risk of neourological damage in the unimmunised however the additional risk would be very small and the evidence is very sparse indeed. This is one area that needs more research.

I believe that we do not know enough about the long term effects of vaccines on our children. In fact I think there is enough evidence to suggest that vaccines may be weakening the immune system and exposing our children to greater risks associated with SIDS, Asthma, and bacterial infections.

The only longer term study (not intended as a review of the pertussis vaccine) has revealed alarming relationships between pertussis vaccines and asthma. What would be the outcome of a more focussed study comparing key health indicators for immunised and unimmunised children and why hasn't this fundamental study been carried out?

I also think that many serious adverse reactions to vaccines go unreported. Which doctor will advocate vaccines to their patient and then 2 weeks later undermine his position completely by admitting that a reaction such as SIDS or a neurological reaction could have been associated with the vaccine?

I think a pro-choice environment would cause a vaccine coverage rate of between 60% and 70% that is not significantly different from the current estimated immunisation rate. I believe that reverting to an environment of responsible choice will not increase notification or incidence rates significantly in Australia.

The treatment of pertussis consists of bed rest, adequate nutrition, and adequate amounts of fluid. Erythromycin may be prescribed to reduce transmission or to control secondary infection. Hospitalisation may be necessary for infants and children with severe or prolonged paroxysms and for those with dehydration or other complications. Oxygen may be needed to relieve dyspnea and cyanosis; intravenous therapy may be necessary when prolonged vomiting interferes with adequate nutrition. Intubation is rarely necessary but may be lifesaving in infants if the thick mucus cannot be easily suctioned from the air passages.

I do not believe that the risk of a child dying from pertussis is reduced significantly by vaccination. I also believe that issues such as SIDS are impacted by the health of a child's immune system. If a child's immune system is compromised in any way during the crucial first 12 months then I think that this does not bode well for its short and long term health prospects.

While the risk of an unimmunised child contracting pertussis can never be eradicated, it can be reduced. By being aware of 4 yearly cycles and the seasonality of the disease we can be vigilant for signs of local outbreaks. If your child is significant contact then antibiotic prophylaxis may be employed to reduce the infectious period and to reduce the symptoms to the extent that the outward signs of the disease can be avoided all together if administered early enough.

Pertussis vaccines and other related pharmaceutical products are big business, currently dominated in Australia by CSL and SmithKline Beecham.

The PBAC considers comparative cost-effectiveness and makes recommendations to the Minister based on assumptions and empirical models developed by the potential vaccine supplier. I think this is a conflict of interests.

These studies are then protected by commercial confidentiality arrangements. So we (the general public) are unable to gain access to the analysis that justifies vaccinating our children.

If the government was doing its job correctly there would be no need for vaccine awareness groups.

18

WHAT'S THE RIGHT ANSWER?

I was hoping to find a black and white answer on whether or not the risks of whooping cough outweigh the risks associated with the vaccine. Unfortunately there isn't a yes or no answer. We have reviewed as much literature as we can (or want to) on the subject and my wife and I made the decision not to vaccinate our four children (the most precious things in the world to us) against whooping cough.

Our children were born full term, breastfed and we avoided childcare for the first 12 months. We decided there were plenty of other things we could be doing to maximise the health of our children (eg. diet, exercise, love, respect, to name a few). Intuitively we felt that the risks with the disease did not outweigh the health risks associated with the vaccine.

Our children are now 7, 7, 10 and 13, and none of them have been injected with any of the vaccines. I can tell you that every cough has been more stressful than it should be. The 10 year old has mild asthma which is continuing to fade with age. Each child endures colds and ailments from time to time although their general health seems to be well above average. They have all had chicken pox and continue to be exposed to all sorts of childhood diseases at school and in the community.

I took the following quote from one of Hugh Mackay's books;

" ..... our insecurities and uncertainties stimulate our desire for security and certainty: we are almost instinctively attracted to the confident voice; the strong leader; the person who seems to know what ought to be done. Moral vacuums yearn to be filled.

The danger, at such times, is that we might settle for an easy certainty, when uncertainty might be more appropriate; we might take refuge in the security of a simple view of the world, when the world might actually be more complex than we would wish."(30)

Then another quote from a lady called Susan Maushart:

"Among the many cultural contradictions about mothering that we harbour, none is more harmful than this: that we believe on the one hand that for every parenting problem there exists some technical or technological solution; and on the other that mothering is a fundamentally instinctive, intuitive process. It's a problem that's been with us ever since Spock."(31)

Perhaps it is time to shift the tide a little.

GLOSSARY (all taken from 38)

Acellular
> without cells.

Aerosols
> nebulised particles suspended in a gas or air.

Anaphylaxis
> an exaggerated, life-threatening hypersensitivity reaction to a previously encountered antigen.

Antibiotic
> pertaining to the ability to destroy or interfere with the development of a living organism. An antimicrobial agent, derived from cultures of a micro-organism or produced semisynthetically, used to treat infections.

Antibody
> an immunoglobulin produced by lymphocytes in response to bacteria, viruses, or other antigenic substances. An antibody is specific to an antigen.

Antigen
> a substance, usually a protein, which the body recognises as foreign and that can evoke an immune response.

Antihistamines
> any substance capable of reducing the physiologic and pharmacologic effects of histamine, including a wide variety of drugs that block histamine receptors.

Antipyretics
> pertaining to a substance or procedure that reduces fever. A common antipyretic agent is aspirin.

Apnoea
> an absence of spontaneous respiration.

Artificial Ventilation
> the process of supporting respiration by manual or mechanical means when normal breathing is inefficient or has stopped.

Aspiration
> the act of taking a breath, inhaling OR the act of withdrawing a fluid, such as mucus or serum, from the body by a suction device.

Assay
> the analysis of the purity or effectiveness of drugs and other biologic substances, including laboratory or clinical observations.

Asthma
> a respiratory disorder characterised by recurring episodes of paroxysmal dyspnea, wheezing on expiration/inspiration caused by constriction of the bronchi, coughing, and viscous mucoid bronchial secretions.

Attenuated
> pertaining to the dilution of a solution or the reduction in virulence or toxicity of a micro-organism or a drug by weakening it.

Bacteria
> the small unicellular micro-organisms of the class Schizomycetes. The genera vary morphologically (in size and shape), being spheric (cocci), rod-shaped (bacilli), spiral (spirochetes), or comma-shaped (vibrios). The nature, severity, and outcome of any infection caused by a bacterium are characteristic of that species.

Bacterial Protein
> a protein produced by a bacterium.

Bordetella
> [Jules J.B.V. Bordet, Belgian bacteriologist, 1870-1961], a genus of gram-negative coccobacilli, some species of which are pathogens of the respiratory tract of humans.

Bronchitis
>   an acute or chronic inflammation of the mucous membranes of the tracheobronchial tree.

Bronchodilators
>   a substance, especially a drug, which relaxes contractions of the smooth muscle of the bronchioles to improve ventilation to the lungs.

Catarrhal
>   pertaining to catarrh, or discharge from an inflamed mucous membrane.

Chemotherapeutic agent
>   a chemical agent used to treat diseases.

Clinical
>   pertaining to materials or equipment used in the care of a sick person.

Coccobacilli
>   spherical bacterial cell.

Coma
>   a state of profound unconsciousness, characterised by the absence of spontaneous eye openings, response to painful stimuli, and vocalisation. The person cannot be aroused.

Congenital Anomoly
>   any abnormality present at birth, particularly a structural one, which may be inherited genetically, acquired during gestation.

Conjunctiva
>   the membrane lining the inner surfaces of the eyelids and anterior part of the sclera.

Conjunctival Hemorrhage
>   bleeding of the conjunctiva.

Conjunctivitis

inflammation of the conjunctiva, caused by bacterial or viral infection, allergy, or environmental factors.

Convalescence

the period of recovery after an illness.

Convulsions

see seizure.

Corticostaroids

any one of the natural or synthetic hormones elaborated by the adrenal cortex that influence or control key processes of the body.

Culture

a laboratory test involving the cultivation of micro-organisms or cells in a special growth medium.

Cyanosis

bluish discoloration of the skin and mucous membranes caused by an excess or deoxygenated hemoglobin in the blood.

Dehydration

excessive loss of water from body tissues. Dehydration is accompanied by a disturbance in the balance of essential electrolytes, particularly sodium, potassium, and chloride. It may follow prolonged fever, diarrhoea, vomiting and any other condition in which there is a rapid depletion of body fluids. It is of particular concern among infants and young children, because their electrolyte balance is normally precarious. Normal fluid volume and balanced electrolyte values are the primary goals of therapy.

Diagnosis

identification of a disease or condition by scientific evaluation of physical signs, symptoms, history, laboratory test results, and procedures.

Diptheria

an acute contagious disease caused by the bacterium Corynebacterium diphtheriae. It is characterized by the production of systemic toxin and a false membrane lining of the mucous membrane of the throat.

Dyspnea

a distressful sensation of uncomfortable breathing that may be caused by certain heart conditions, strenuous exercise, or anxiety.

Efficacy

the maximum ability of a drug or treatment to produce a result, regardless of dosage.

Encephalitis

an inflammatory condition of the brain.

Encephalopathy

any abnormal condition of the structure or function of brain tissues, especially chronic, destructive, or degenerative conditions.

Endemics

the expected or "normal" incidence indigenous to a geographic area or population.

Endotoxin

a toxin contained in the cell walls of some micro-organisms, especially gram-negative bacteria, that is released when the bacterium dies and is broken down in the body. Fever, chills, shock and a variety of other symptoms result, depending on the particular organism and the condition of the infected person.

Endotracheal Intubation

the management of the patient with an airway catheter inserted through the mouth or nose into the trachea. An endotracheal tube may be used to maintain a patent airway, to prevent aspiration of material from the digestive tract in the unconscious or paralysed patient, to minister positive-pressure ventilation that cannot be given effectively by a mask. Endotracheal tubes may be made of rubber or plastic and usually have an inflatable cuff to maintain a closed system with the ventilator.

Epidemic

affecting a significantly large number of people at the same time. A disease or event whose incidence is beyond what is expected.

Epidemiology (/ical)

the study of the determinants of disease events in populations.

Epithelium

the covering of the internal and external organs of the body and the lining of the vessels, body cavities, glands, and organs.

Erythromycin

an antibacterial antibiotic.

Expectorants

pertaining to a substance that promotes the ejection of mucus or other exudates from the lung, bronchi, and trachea.

Fimbria

any structure that forms a border or edge or that resembles a fringe.

Fit

a paroxysm or seizure.

Frenulum of tongue

a longitudinal fold of mucous membrane connecting the floor of the mouth to the underside of the tongue in midline.

Frenum

a restraining portion or structure.

Gastrointestinal

pertaining to the organs of the gastrointestinal tract, from mouth to anus.

Gestation

the period from the fertilization of the ovum until birth.

Gram Negative

[Hans C.J. Gram, Danish physician, 1853-1938], having the pink colour of the counterstain used in Gram's method of staining micro-organisms. This property is a primary method of characterising organisms in microbiology.

Haemophilus Influenzae b (HIB)

a small gram-negative nonmotile parasitic bacterium that occurs in two forms, encapsulated and nonencapsulated, and in six types, a, b, c, d, e, and f. Almost all infections are caused by encapsulated type b organisms.

Haemophilus influenzae

is found in the throats of 30% of healthy, normal people. In children and in debilitated older people, severe destructive inflammation of the larynx, trachea, and bronchi may result from infection. Sub-acute bacterial endocarditis and purulent meningitis also may be caused by it.

Hepatitis B

a form of viral hepatitis caused by the hepatitis B virus (HBV). The virus is transmitted in contaminated serum in blood transfusion, by sexual contact with an infected person, or by the use of contaminated needles and instruments.

Hernia

protrusion of an organ through an abnormal opening in the muscle wall of the cavity that surrounds it.

Histamine

a compound found in all cells, produced by the breakdown of histidine. It is released in allergic inflammatory reactions and causes dilation of capillaries, decrease in blood pressure, increase in secretion of gastric juice, and constriction of smooth muscles of the bronchi and uterus.

Hypertension

a common, often asymptomatic disorder characterized by elevated blood pressure persistently exceeding 140/90 mm Hg.

Hypo-

prefix meaning under, below, beneath, deficient.

Hypoglycemia

a less than normal amount of glucose in the blood, usually caused by administration of too much insulin, excessive secretion of insulin by the islet cells of the pancreas, or dietary deficiency. The condition may cause weakness, headache, hunder, visual disturbances, ataxia, anxiety, personality changes, and if untreated, delirium, coma, and death. The treatment is the administration of glucose in orange juice (or other fluids) by mouth if the person is conscious or in an intravenous glucose solution if the person is unconscious.

Hypotonic

pertaining to a lower or lessened tone or tension that may involve any body structure.

Hypoxia

inadequate oxygen at the cellular level, characterised by tachycardia, hypertension, peripheral vasoconstriction, dizziness, and mental confusion. Treatment may include cardiotonic and respiratory stimulant drugs, oxygen therapy, mechanical ventilation, and frequent analysis of blood gases.

Immunisation

a process by which resistance to an infectious disease is induced or augmented.

Immunogenicity
> the ability of an antigen to induce a specific immune response.

Immunoglobulin
> any of five structurally and antigenically distinct antibodies present in the serum and external secretions of the body. In response to specific antigens, immunoglobulins are formed in the bone marrow, spleen, and all lymphoid tissue of the body except the thymus. Kinds of immunoglobulins (Igs) are IgA, IgD, IgE, IgG, and IgM.

Infection
> the invasion of the body by pathogenic micro-organisms the reproduce and multiply, causing disease by local cellular injury, secretion of a toxin, or antigen-antibody reaction in the host.

Infectious
> capable of causing an infection.

Inguinal
> pertaining to the groin.

Intravenous Therapy
> the administration of fluids or drugs, or both, into the general circulation through a puncture of a vein.

Intubation
> the insertion of a tube through the mouth or nose into the trachea to ensure a patent airway for the delivery of anaesthetic gases and oxygen or both.

Invasive Bacterial Infections
> bacterial infections characterised by a tendency to spread, infiltrate, and intrude.

Isolates
> to derive from any source a pure culture of a micro-organism.

Laryngeal
> pertaining to the larynx.

Larynx

the organ of voice that is part of the air passage connecting the pharynx with the trachea. It accounts for a large bump in the neck called the Adam's apple and is larger in men than in women.

Lymphocyte

small agranulocytic leukocytes originating from fetal stem cells and developing in the bone marrow. Lymphocytes normally comprise 25% of the total white blood cell count but increase in number in response to infection. Two forms occur: B cells and T cells. B cells circulate in an immature form and synthesize antibodies for insertion into their own cytoplasmic membranes. When an immature B cell is exposed to a specific antigen, the cell is activated, travelling to the spleen or to the lymph nodes, differentiating, and rapidly producing plasma cells and memory cells. Plasma cells synthesize and secrete antibody. Memory cells do not secrete antibody but, on re-exposure to the specific antigen, develop into antibody secreting plasma cells. T cells are lymphocytes that have circulated through the thymus gland and have differentiated to become thymocytes. When exposed to an antigen they divide rapidly and produce large numbers of new T cells sensitised to that antigen. Some T cells are often called 'killer cells' because they secrete immunologically essential chemical compounds and assist B cells in destroying foreign protein. T cells also appear to play a significant role in the body's resistance to the proliferation of cancer cells.

Macrolide

any group of antibiotics produced by actinomycetes. They include erythromycin and troleandomycin. Macrolides are generally used against gram-positive bacteria and in patients allergic to penicillins.

Malaise

a vague uneasy feeling of body weakness, distress, or discomfort, often marking the onset of and persisting throughout a disease.

Malnutrition
> any disorder of nutrition. It may result from an unbalanced, insufficient, or excessive diet or from impaired absorption, assimilation, or use of foods.

Meningitus
> any infection or inflammation of the membranes covering the brain and spinal cord.

Meningococcus (Meningococcol, *adj*)
> a bacterium of the genus Neisseria meningitidis, nonmotile gram-negative diplococcus, frequently found in the nasopharynx of asymptomatic carriers, which may cause septicemia or epidemic cerebrospinal meningitis. Meningococcal infections are not highly communicable; however, crowded conditions, such as may be found in army camps, concentrate the number of carriers and reduce individual resistance to the organism. Hemorrhagic skin lesions are significant clues to the diagnoses. Stained smears of these lesions or of cerebrospinal fluid must be examined quickly because meningococci are fragile and lyse readily. Early treatment with an appropriate antibiotic such as penicillin G is essential for cure.

Mortality
> the death rate, which reflects the number of deaths per unit of population in any specific region, age group, disease, or other classification.

Mucus
> the viscous, slippery secretions of mucous membranes and glands, containing mucin, white blood cells, water, inorganic salts and exfoliated cells.

Nasopharyngeal
> pertaining to the cavity of the nose and the nasal parts of the pharynx.

Neurological (neurology)
> the field of medicine that deals with the nervous system and its disorders.

Neurotoxic
> having a poisonous effect on nerves and nerve cells, such as when ingested lead degenerates peripheral nerves.

Organisms
> an individual living animal or plant able to carry on life functions through mutually dependent systems and organs.

Oropharyngeal (oropharynx)
> one of the three anatomic divisions of the pharynx. It extends behind the mouth from the soft palate above to the level of the hyoid bone below and contains the palatine and lingual tonsils.

Osteoporosis
> a disorder characterized by abnormal loss of bone density.

Otitis Media
> inflammation or infection of the ear.

Pallor
> an unnatural paleness or absence of colour in the skin.

Parapertussis
> an acute bacterial respiratory infection caused by Bordetalla parapertussis, having symptoms closely resembling those of pertussis. It is possible to be infected with both B. parapertussis and B. pertussis at the same time.

Paroxysm
> a convulsion, fit, seizure, or spasm.

Pathogen
> any micro-organism capable of producing disease.

Pathophysiology
> the study of the biologic and physical manifestations of disease as they correlate with the underlying abnormalities and physiologic disturbances. Pathophysiology does not deal directly with the treatment of disease; rather, it explains the processes within the body that result in the signs and symptoms of a disease.

Perinatal

pertaining to the time and process of giving birth of being born.

Pertussis

an acute highly contagious respiratory disease characterized by paroxysmal coughing that ends in a loud whooping inspiration.

Pneumonia

an acute inflammation of the lung.

Premature

not fully developed or mature, occurring before the appropriate or usual time.

Prodromal

pertaining to early symptoms that may mark onset of a disease.

Prolapse

the falling, sinking, or sliding of an organ from its normal position or location in the body.

Prophylactic

preventing the spread of disease, an agent that prevents the spread of disease.

Rectal (Rectum)

the part of the large intestine, about 12cm long, continuous with the descending sigmoid colon, proximal to the anal canal.

Retrospective Study

a study in which a search is made for a relationship between one (usually current) phenomenon or condition and another that occurred in the past.

Reyes Syndrome

[Ralph D. K. Reye, twentieth century Australian pathologist] a combination of acute encephalopathy and fatty infiltration of the internal organs that may follow acute viral infections.

Rhinorrhea

the free discharge of a thin watery nasal fluid.

Rubella

a contagious viral disease characterised by fever, symptoms of a mild upper respiratory tract infection, lymph node enlargement, arthralgia, and a diffuse fine red maculopapular rash.

Secretions

the release of chemical substances manufactured by cells of glandular organs.

Sedatives

pertaining to a substance, procedure of measure that has a calming effect.

Seizure

a hyperexcitation of neurons in the brain leading to a sudden, violent involuntary series of contractions of a group of muscles. It may be paroxysmal and episodic, as in a seizure disorder, or transient and acute, as after a head concussion.

Sequelae

any abnormal condition that follows and is the result of a disease, treatment, or injury.

Serologic

pertaining to the branch of medicine concerned with the study of blood sera.

Serology

the branch of laboratory medicine that studies blood serum for evidence of infection by evaluating antigen-antibody reactions in vitro.

Spasms

an involuntary muscle contraction of sudden onset, such as habit spasms, hiccups, stuttering, or a tic.

Stridor

an abnormal high-pitched musical sound caused by an obstruction in the trachea or larynx.

Sudden Infant Death Syndrome (SIDS)

an unexpected and sudden death of an apparently normal and healthy infant that occurs during sleep and with no physical or autopsic evidence of the disease.

Swab

a stick or clamp for holding absorbent gauze or cotton, used for washing, cleansing, or drying a body surface; for collecting a specimen for laboratory examination; or for applying a topical medication.

Tetanus

an acute, potentially fatal infection of the central nervous system caused by an exotoxin, tetanospasmin, elaborated by an anaerobic bacillus, Clostridium tetani.

Trachea

a nearly cylindric tube in the neck, composed of cartilage and membrane, which extends from the larynx at the level of the sixth cervical vertebra to the fifth thoracic vertebra, where it divides into two bronchi. The trachea conveys air to the lungs, it is about 11 cm long and 2 cm wide. The ventral surface of the tube is covered in the neck by the isthmus of the thyroid gland and various other structures such as the sternothyroideus and sternohyoideus. Dorsally the trachea is in contact with the esophagus. Also called windpipe.

Tracheobronchial tree

an anatomical complex that includes the trachea, bronchi, and bronchial tubes. It conveys air to and from the lungs and is a primary structure in respiration.

Tracheostomy

an opening through the neck into the trachea through which an indwelling tube may be inserted.

Tussive

cough.

Ulcer

a circumscribed, craterlike lesion of the skin or mucous membrane resulting from necrosis that accompanies some inflammation, infectious, or malignant process.

Ulceration

the process of ulcer formation.

Umbilical Hernia

a soft, skin covered protrusion of intestine and omentum through a weakness in the abdominal wall around the umbilicus. It usually closes spontaneously within 1 to 2 years, although large hernias may require surgical closure.

Vaccine

a suspension of attenuated or killed micro-organisms administered intradermally, intramuscularly, orally, or subcutaneously to induce active immunity to infectious disease.

Vaccination

any injection of vaccine administered to induce immunity or to reduce the effects of associated infectious disease.

Virulent

pertaining to a very pathogenic or rapidly progressive condition.

Virus

a minute parasitic micro-organism much smaller than a bacterium that, having no independent metabolic activity, may replicate only within a cell of a living plant or animal host.

BIBLIOGRAPHY

(1)   L Gustafsson, H O Hallander, P Olin, E Reizenstein, J Storsaeter (1996) A Controlled Trial of a Two Component Accelular, A Five Component Accellular, and a Whole-Cell Pertussis Vaccine. N Engl J Med:334:349-55.

(2)

(3)

(4)

(5)   Centers for Disease Control and Prevention. Resurgence of Pertussis - United States, 1993. MMWR Morb Mortality Wkly Rep 1993;42:952-3,959-60.

(6)   H E de Melker, M A E Conyn-van Spaendonk, H C Rumke, J K van Wijngaarden, F R Mooi, J F P Schellekens (1997) Pertussis in the Netherlands: an outbreak despite high levels of Immunization with Whole-Cell Vaccine. Emerging Infectious Diseases: 3(2) April-June.

(7)   V Romanus; R Jonsell; SO Bergquist (1987) Pertussis in Sweden after the cessation of general Immunisation in 1979. Pediatr Infect Dis J Apr;6(4): 364-71.

(8)   Whooping Cough Surveillance - Epidemiology. Brit Med J (1982) Nov; 285: 1583-84.

(9)   Estimated Resident Population by Country of Birth, Age and Sex Australia June 1990 and Preliminary 1991 - Australian Bureau of Statistics. Catalogue No. 3221.0.

(10)   Grob R, M J Crowder, J F Robbins (1981) Effect of Vaccination on Severity and dissemination of Whooping Cough. Brit Med J, 13 June; 282: 1925-28.

(11)   N. McK. Bennet (1973) Whooping Cough In Melbourne. Med J of Aust, Sept 8: 481-487.

(12)   G.T. Stewart (1977) Vaccination Against Whooping Cough. Efficacy versus Risk. Lancet, Jan 29: 234-237.

(13)   J Gillis, T Grattan-Smith, H Kilham (1988) Artificial Ventilation in Severe Pertussis. Arch of Disease in Childhood, 63:364-367.

(14)   B.J. Feery, W.K. Finger, Z Kortus, G.A. Jones (1985) The Incidence and Type of Reactions to Plain and Absorbed DTP Vaccines, Aust. Paediatr. J, 21:91-95.

(15)   D.L. Miller, E.M. Ross, R. Alderslade, M.H. Bellman, N.S.B. Rawson (1981) Pertussis Immunisation and Serious Acute Neurological Illness in Children, Brit Med J, 282:1595-1599.

(16)   C.P. Howson, H.V. Fineberg (1992) Adverse Events Following Pertussis and Rubella Vaccines, JAMA, 267(3):392-396.

(17)   L.J Baraff, W.D. Shields, L. Beckwith, G Strome, S.M. Marcy, J.D. Cherry, C.R. Manclark (1988) Infants and Children with Convulsions and Hypotonic-Hyporesponsive Episode Following DTP Immunisation, Pediatrics, 81: 789-794.

(18)   J.L Carroll, A.S. Siska (1998) SIDS: Counselling Parents to Reduce the Risk, American Family Physician, April 1.

(19)   R.G. Carpenter, A Gardner (1979) Identification of some infants at immediate risk of Dying Unexpectedly and Justifying Intensive Study, Lancet, Aug. 18: 343-346.

(20)   V. Scheibner (1993) Vaccination: 100 Years of Orthodox Research shows that Vaccines Represent a Medical Assault on the Immune System, Pub. by Viera Scheibner, Blackheath, NSW.

(21)   G.T. Stewart (1979) Deaths of Infants after Triple Vaccine, Lancet, Aug 18: 354-355.

(22)   M.R. Griffin, W.A. Ray, J.R. Livengood, W Schaffner (1988) Risk of Sudden Infant Death Syndrome after Immunization with the DTP Vaccine, N Engl J Med; 319:618-23.

(23)   M.R. Odent (1994) Pertussis Vaccination and Asthma: Is There a Link?, JAMA, (272)8:593.

(24)   J Storsaeter, P Olin, B Renemar, T Lagergard, R Norberg, V Romanus, M Tiru (1988) Mortality and morbidity from invasive bacterial infections during a clinical trial of acellular pertussis vaccines in Sweden, Pediatr Infect Dis J, 7:637-645.

(25)   R H Michaels (1971) Increase in Influenzal Meningitis, N Engl J Med, 285(12):666-667.

(26)   K A V Cartwright, J M Stuart, N D Noah (1986) An Outbreak of Meningococcal Disease in Gloucestershire, Lancet, Sept 6:558-561.

(27)

(28)   N Wortis, P M Strebel, M Wharton, B Bardenheier, I R B Hardy (1996) Pertussis Deaths: Report of 23 Cases in the United States, 1992 and 1993, Pediatrics, 97(5):607-612.

(29)   M J Goldachre, R I Harris (1981), Hospital admissions for Whooping Cough in the Oxford Region, 1974-9, Brit Med J, 282: 106-107.

(30)   H Mackay(1997), Generations: Baby Boomer, their parents & their children. Pub. by Pan Macmillan Australia Pty Ltd. Sydney, Australia:190.

(31)   S Maushart(1997), The Mask of Motherhood. Pub. by Random House Australia Pty Ltd. Milsons Point, Australia: 191.

(32)   P. McIntyre, J. Amin, H. Gidding, et al (June 2000), Vaccine Preventable Diseases and Vaccination Coverage in Australia, 1993-1998. Pub. By National Centre for Immunisation Research and Surveillance of Vaccine Preventable Diseases (NCIRS) as a Supplement for Communicable Diseases Network Australia.

(33)   Pertussis Working Party (PWP) (1997), Guidelines for the control of Pertussis in Australia – Technical Report Series No. 1, Published by the Commonwealth Department of Health and Family Services, Publication Identification Number: 2337, for Communicable Diseases Network Australia New Zealand.

(34)   P. McIntyre, J. Forrest, T. Heath, M Burgess, B Harvey (1998), Pertussis vaccines: past, present and future in Australia, Comm Dis Intell;22:125-132.

(35)   National Health and Medical Research Council (1997), The Australian Immunisation Handbook, Published by the Australian Government Publishing Service.

(36)   D. Jenkinson (February 1988), Duration of effectiveness of pertussis vaccine a 10 year community study, Brit Med J, 296: 612-614.

(37)   E. Gangarosa, A Galazka, C Wolfe, L Phillips, R Gangarosa, E. Miller, R. Chen (1998), Impact of anti-vaccine movements on pertussis control: the untold story, Lancet, 351: 356-61.

(38)   K. Anderson (Editor)(1999), Mosby's medical, nursing, & allied health dictionary. Revision 5. Pub. by Mosby, Inc., USA.

(39)   Causes of Death Australia 1998 - Australian Bureau of Statistics. Catalogue No. 3303.0.

(40)   KM Farizo, SL Cochi, ER Zell, EW Brink, SG Wassilak, PA Patriarca (1992), Epidemiological features of pertussis in the United States, 1980 - 1989, Clinical Infectious Diseases, 14: 708-19.

(41)   Kaplan (1993), Current Therapy in Pediatric Infectious Disease. 3rd Edition.

(42)   J Taranger (1982), Mild Clinical Course of Pertussis in Swedish Infants of Today, Lancet, June 12:1360.

(43)   Neenig ME, Shinefield HR, Edwards KM, Black SB, Fireman BH (1996), Prevalence and incidence of Adult Pertussis in an Urban Population, JAMA, June 5; 275(21): 1672-4.

NOTES

NOTES